SOFT

MAGNETISM

Fundamentals for
Powder Metallurgy and Metal Injection Molding

SOFT MAGNETISM

Fundamentals for
Powder Metallurgy and Metal Injection Molding

by

Chaman Lall, Ph.D.

Vice-President and Technical Director
Midwest Sintered Products Corporation
Riverdale, Illinois

METAL POWDER INDUSTRIES FEDERATION
Princeton, New Jersey

Library of Congress Cataloging-in-Publication Data

Lall, Chaman.
    Soft magnetism : fundamentals for powder metallurgy and metal injection molding / by Chaman Lall.
        p. cm. — (Monographs in P/M series, ISSN 1061-6071 ; no. 2)
    Includes bibliographical references and index.
    ISBN 1-878954-17-2
    1. Powder metallurgy. 2. Injection molding of metals. 3. Metals-Magnetic properties. I. Title. II. Series.
TN695.L35 1992
671.3'7—dc20   92-19313
                                                                    CIP

Copyright 1992
METAL POWDER INDUSTRIES FEDERATION
105 College Road East
Princeton, New Jersey 08540-6692 USA

ISSN 1061-6071
ISBN 1-878954-17-2
Library of Congress Catalog Card No. 92-19313

Printed in the United States of America

CONTENTS

CONTENTS *Continued*

In remembrance of
The One
who taught me all
I ever really needed to know.

CHAMAN LALL

Chaman Lall is Vice-President and Technical Director at Midwest Sintered Products Corporation, a powder metal parts producer in the Riverdale suburb of Chicago, Illinois, U.S.A. His technical interests include process and materials research on high-performance soft magnetic materials, stainless steels, and structural steels.

Chaman joined this facility in 1989, after nearly a decade of successful research and management activities with E.I. du Pont de Nemours and Company, and the P/M group within its Remington Arms Company. These activities included the development of high performance alloys and production processes for metal and ceramic injection molding. He also performed fundamental research and development at the University of Pennsylvania (1976-1977) and Drexel University (1978-1979). Chaman studied "Physical Metallurgy and the Science of Materials" at the University of Birmingham (England, UK), and received his B.Sc. degree in this subject in 1972 and a Ph.D. degree in 1976.

PREFACE

The Iron Pillar[1] in Delhi, India, is often mentioned as one of the finest early examples of an article made by the process we call powder metallurgy (P/M). This particular example illustrates two aspects of powder metallurgy that prevail even today. The first is the degree of excellence to which a powder metal product can serve in a particular application, and the second is the "black art" of processing that allows one to make such an outstanding product.

Articles made by powder metallurgy are often perceived to have very poor corrosion resistance. Yet this Iron Pillar is over 1,500 years old and has held up very well against the elements for all these centuries. What is also remarkable is that the capability to handle such a large iron article (24 feet tall and weighing more than 6 tons) would not be possible in Europe until the 19th century. It is now believed that the silicon in the iron ore found in local quarries and used to make the pillar resulted in a dual benefit. First, the high silicon content enabled the early practitioners to bond the lumps or granules at low sintering temperatures. Second, the silicon in the form of a complex oxide coating turned out to be an excellent barrier to corrosion. It is unlikely that the "powder metallurgists" of that era fully understood the science of their process.

A similar sense of awe and confusion appears to surround the parts produced by powder metallurgy for sophisticated magnetic applications. Magnetism is regarded as a mystical phenomenon because of our inability to touch and feel it. Much in the same manner as gravity, one can only observe the effects of magnetism. Magnetism is purported[2] to be the force behind exotic devices that can improve the taste of wine, soften water, make hair silkier, improve automobile gas mileage, and reduce exhaust emissions!

The fact is that both magnetism and powder metallurgy are understood to the extent that each can be practiced as a science. It is the purpose of this monograph to bring these two subjects together in a sufficiently coherent manner so that both producers and users of powder metal components can discuss and develop magnetic applications in a systematic manner.

Powder metallurgy has developed into a scientifically understood fabrication technology within the last half of the twentieth century. Much of the focus has been on developing products for structural applications, so that the majority of components in use today utilize the mechanical attributes of the P/M offering.

---

1. "New Delhi: India's Mirror", *National Geographic,* Vol. 167, No. 4, April, 1985.
2. "Unusual Claims About Magnets Attract Suits", *Wall Street Journal,* p. B1, September 24, 1991.

A potentially strong growth market for P/M is one where the magnetic properties of a component are functionally important. The products for this general market can be divided into two categories; soft and hard magnetic devices. These quite distinctly separate groups, while falling in the general field of magnetism, have totally different objectives during the manufacture of the product. Soft magnetic components are produced with the goal of obtaining a defect-free microstructure, with as large a grain size as possible. In contrast, hard or permanent magnetic components should have as many precipitates, inclusions, etc., as possible.

The fundamentals of the powder metallurgy process are described, with a view to relating how the process variables influence the soft magnetic performance of the product. A logical extension of the powder metal process is the newer technology referred to as metal injection molding (MIM). Conventional P/M is limited to producing shapes that can be formed by the vertical motion of tooling members, inside a die that creates the peripheral features of the part. The MIM technology enables features to be formed by the motion of tooling members normal to the vertical direction. In this way, the MIM forming method complements the conventional P/M process in that another dimension can be freely designed by the user. A truly three-dimensional shape forming technology is in the offering. Furthermore, the MIM process has the potential to process a wider range of alloys and produce a more uniform product, than P/M.

The primary emphasis in this text is to promote an understanding of the magnetism phenomenon and describe the important parameters that define the magnetic characteristics of a given material. Some selected test methods for the evaluation of magnetic properties are briefly described. Furthermore, the relationship of the measured parameters to the performance of soft magnetic devices is discussed.

The properties of materials manufactured by powder metallurgy and metal injection molding are presented and, in addition, compared to their wrought counterparts. Some example devices are described to illustrate the kinds of applications that require soft magnetic properties. This book is complemented with several appendices that provide guidance on terminology and standards that are in current use.

No attempt has been made to derive well-known formulae that may be found in standard text books on physics. The primary focus is on providing a qualitative and practical understanding of the phenomena under discussion. Wherever possible, the International System of Units (SI) has been used. However, the c.g.s. system of units is fairly widely-accepted and preferred for magnetism in the U.S.A. and many other parts of the world. For this reason, values using the c.g.s. or other "customary" units have been included in parentheses.

This book is primarily intended for those individuals responsible for the optimization of the manufacturing process as well as those engineers designing and specifying soft magnetic components. The pursuit of education of the engineering and marketing communities has been the driving force behind this book.

I would like to express my appreciation to many of my friends and colleagues for their support. In particular, I wish to thank Mr. H. C. (Hal) Munson and Mr. Louis W. Baum, of Remington Arms Company, and Professor C. D. (Chad) Graham, of the University of Pennsylvania, for introducing me to this subject and for the motivation to pursue this intriguing and fascinating field. I also wish to thank all of my associates who supplied data or examples for this monograph; each is recognized at the appropriate point in the text. An endeavor such as this would have been practically impossible without the support and encouragement from my wife Samitra, to whom I am very grateful. Finally, my thanks to all the unsung heroes at MPIF headquarters for their assistance in completing this monograph.

Chaman Lall
Chicago, Illinois, USA
September, 1992

I.

FUNDAMENTALS
OF MAGNETISM

The fundamental basis of magnetism is developed with regard to the effect of a moving charged particle and the influence of material structure. The majority of this initial chapter is devoted to describing soft magnetism and the associated phenomena. Short sections on permanent magnetism and some selected measurement techniques serve to complete this introduction to basic magnetism.

1.1 Introduction

The fascinating aspect of magnetism is its ability to generate forces between objects that are not in physical contact with each other. This invisible force is responsible for our inability to push together two powerful magnets that are oriented so as to repel each other. The same force enables a permanent magnet to pick up a pile of paper clips or nails that were previously unattracted to each other.

Magnetic forces are just one example of several forces that can cause interactions between matter that is not in physical contact. The concept of forces and force fields is developed first, in order to provide a framework to describe magnetic interactions. Initial discussions focus on gravitational, electrostatic and magnetic fields, since an understanding of these subjects serves to demonstrate the similarities (and differences) between them. This is followed by a discussion of how a magnetic field created by the flow of electric current interacts with magnetic materials. Some terminology and the parameters used to characterize magnetic properties are introduced at this point.

1.2 Force Fields

While the exact mechanisms that enable forces to interact across empty space are not known, scientists have introduced the concept of "force fields". A force field can be described as a "distortion" or "perturbation" of space. This distortion may be caused by a mass, a charge, or a current element in the case of gravitational, electrostatic, or magnetic fields, respectively. A specific type of force field will only interact with a given element; a gravitational field with a mass, an electrostatic field with a charge, and a magnetic field with a current element. An illustration of this is a magnetic compass needle that is unaffected either by the earth's gravitational field or the electrostatic field of a charged glass rod that is brought near it. A force field may exist in space, even though there is nothing for it to act upon.

A mass **m** in a gravitational field **g** will generate a gravitational force F_G as follows:

$$F_G = m\, g \qquad\qquad [1.1]$$

Note that this is simply a specific example of **Newton's Second Law of Motion,** which **defines** force as a product of mass and acceleration. In honor of Sir Isaac Newton (1642-1727), the unit of force in the SI system (kg·m/s^2) is called a newton (N); this force will cause a 1 kilogram mass to accelerate at a rate of 1 m/s^2. From Equation [1.1], the gravitational field, **g**, is the gravitational force per unit mass (i.e., newton/kg).

The gravitational field, **g**, varies with the distance **R** from a mass **m** as follows:

$$g = \frac{m}{R^2} \qquad\qquad [1.2]$$

The gravitational force of attraction between two masses at a distance **R** from each other is given by **Newton's Universal Law of Gravitation:**

$$F_G = \frac{Gm_1 m_2}{R^2} \qquad\qquad [1.3]$$

The proportionality constant G is called the **"gravitational constant"** and has the value 6.67 x 10^{-11} m^3/s^2·kg in the SI system (6.67 x 10^{-8} cm^3/s^2·g, in the c.g.s. system). This same constant is applicable between any two bodies in the universe.

In a similar manner to Equation [1.1], a charge **q**, in an electrostatic field **E**, will generate an electrostatic force F_E:

$$F_E = q\, E \qquad\qquad [1.4]$$

2

The electrostatic field is therefore the electrostatic force per unit charge.

Analogous to Equation [1.2], the electrostatic field, **E**, at distance **R** from a point charge **q** is:

$$E = \frac{q}{R^2} \qquad [1.5]$$

According to **Coulomb's Law**, the force exerted by one charged particle on another charged particle is inversely proportional to the square of the distance between them:

$$F_E = \frac{k\, q_1 q_2}{R^2} \qquad [1.6]$$

This is entirely analogous to Equation [1.3] for gravity, except a +/- term has to be introduced to reflect the effect of like and unlike charges. The term **k** is simply a proportionality constant and has the value 8.99×10^9 Newtons·m^2/ Coulomb^2 (1 Dyne·cm2/statocoulombs^2). The force F_E has the units Coulombs^2/m^2 (Coulombs^2/cm^2, in c.g.s.).

Both of these forces are similar, in that the gravitational and electrostatic forces between two bodies are inversely proportional to the square of the distance between their centers. Furthermore, the direction of the forces is along the line joining the centers of the two bodies.

Contrary to gravitational forces, electrostatic and magnetic forces can be either attractive or repulsive, depending on the polarity of the charge or the direction of the current element. One important consequence of this is that the effects of electrostatic and magnetic forces can be cancelled by in-kind forces. Therefore, one can shield against electrostatic and magnetic forces, but not against gravitational ones.

Analogous to the conservation of mass is the law of conservation of charge. This states that a charge can neither be created nor destroyed; the development of charged particles is simply the net transfer of charges from one body to another. When equal amounts of positive and negative charges are present, the body is said to have a neutral charge. Such a body will stay neutral (i.e., have zero net charge) until one adds or removes positive or negative charges.

In the case of gravitational and electrostatic forces, the two bodies or charged particles may be completely at rest and one will still experience these forces. To develop a magnetic field and, therefore a magnetic force, the additional concept of a **moving charge** needs to be introduced. A moving charge or groups of charges is, of course, an electric current. This is simply electricity.

The basis of all magnetic phenomena is the movement of, and

interaction between, charged particles. This very fundamental and important concept should be understood to appreciate the discussions to follow. The fact that there is any relationship between magnetism and electricity, was not realized until the early nineteenth century experiments of Hans Christian Oersted (1777-1851) and Andre Marie Ampere (1775-1836). The former was the first to demonstrate the interaction of the two, by the deflection of a compass needle placed over a wire carrying an electric current. Ampere showed that forces exist between two wires through which electricity is flowing. Furthermore, he was able to create magnetic dipoles by forming a loop in a current-carrying wire. This was the very beginning of the science that is now termed **electromagnetism**.

The discussions to follow will be confined to concepts presented by classical physics; the refinements offered by quantum physics, although more accurate, are beyond the scope of this text. The duality of the electron having the properties of both a particle and a wave is one such example of a possible source of confusion. The idea of an electron as being a negatively charged particle will be retained, for the purposes of this text.

In classical physics, the electron is thought to spin on its own axis, similar to the rotation of the earth. In addition, the electron also rotates around the nucleus, much as the earth rotates around the sun. One can consider the so called "electron spins" and electron rotation around the atomic nucleus (which contains the neutrons and the positively charged protons) as the origin for the magnetic phenomena at the atomic level. The motion of these charged particles at the atomic level is responsible for the properties exhibited by a solid bar magnet, even though the object may appear to be totally stationary at the macro level.

1.3 Magnetic Field, H

Analogous to Equations [1.1] and [1.4], a magnetic pole **p** in a magnetic field **H** will generate a magnetic force F_M:

$$F_M = p\,H \qquad\qquad [1.7]$$

The magnetic field is therefore the magnetic force per unit pole strength.

Similarly, the magnetic field, **H**, at distance **R** from a magnetic pole **p** is:

$$H = \frac{p}{R^2} \qquad\qquad [1.8]$$

The SI units of magnetic field are amperes/meter (Oersteds).

If p_1 and p_2 represent the strengths of two magnetic poles, the force between them is inversely proportional to the square of the distance between them:

$$F_M = \frac{C\, p_1\, p_2}{R^2} \qquad\qquad [1.9]$$

This is entirely analogous to Equations [1.3] and [1.6], for gravity and electrostatics, respectively. The term **C** is simply a proportionality constant. In honor of Wilhelm Eduard Weber (1804-1891), the SI unit of magnetic force is the weber.

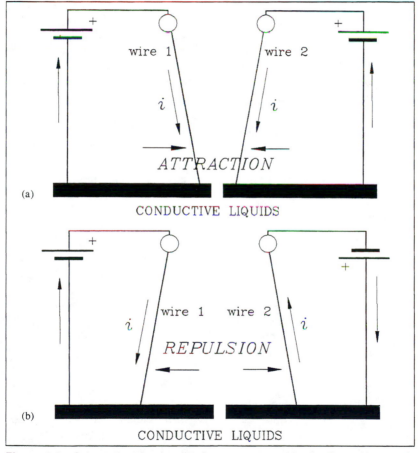

Figure 1.1: Schematic, showing the forces generated by the flow of current in electrical conductors that were initially parallel:
    a) Attractive, with current flow in the same direction.
    b) Repulsive, with current flow in opposite direction.

Consider the experiment depicted in Figure 1.1, where the two wires are initially parallel to each other. With the currents flowing in the same direction, the wires are pulled towards each other, while they are repelled when the electric currents are flowing in opposite directions. This is similar to the observations first made by Ampere in 1820. Clearly, these observations are not caused by electrostatic forces, since each wire has a zero net charge. Furthermore, with electrostatic forces, like charges repel and unlike charges attract, which is in contradiction to the observations. These observations are a result of a new force that can also operate across a void, and is termed magnetism. It is important to note that in the example of the two parallel wires, the resulting force of attraction or repulsion is completely non-existent when the electric current is switched off.

Electric current is the flow of charged particles. Figure 1.2a illustrates the flow of negatively-charged electrons in a wire and the magnetic lines of force that are generated. These magnetic lines of force can be demonstrated by passing a wire through a stiff piece of paper on which some iron filings have been sprinkled. Gently tapping the paper as the current is flowing in the wire, allows the iron filings to orientate themselves and form perfectly circular rings around the wire (Figure 1.2a). This is due to fact that the iron filings tend to line up like miniature compass needles, along the lines of magnetic force.

In determining the direction of the force field, it is important to note that, by convention, the positive direction of current is opposite to the direction of flow of the negatively-charged electrons. This is the direction a positively charged particle would move. Using the **right hand rule,** the thumb is pointed in the direction of the current flow while the fingers of the right hand curve in the direction of the magnetic field. Note that by convention, the arrows on H signify the direction that a north pole would move. Another mental reminder is the **clock rule**, in which the electric current is considered to go into the center of the clock and the observer notes that the rotation of the hands in the clockwise direction depicts the direction of the magnetic field. Note that in reality a single magnetic pole cannot exist. This is in contrast to gravitational and electrostatic fields, where it is possible for a mass or a charged particle, respectively, to exist as a singularity.

A number of excellent text books on physics and magnetism are available for the further detailed study of matter and our universe. The text by Cullity[1] can serve as an outstanding resource on the subject of magnetism and is the basis of the derivations to follow. Another excellent reference is the classical work by Bozorth[2] on ferromagnetism.

In the example of a single straight piece of wire (Figure 1.2a), a current

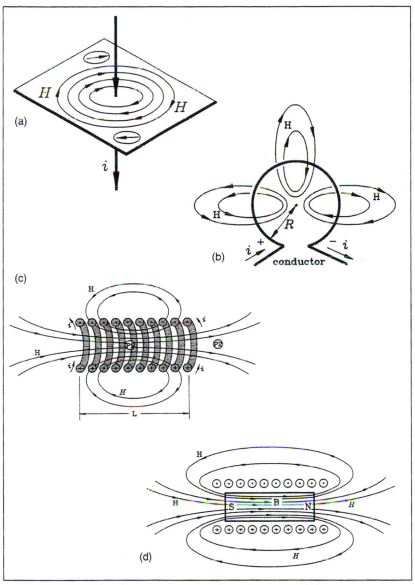

Figure 1.2: Magnetic fields produced by the flow of electric current - schematic.
(a) Magnetic Field around a wire carrying an electrical current. Note the direction of the compass needle at different locations.
(b) Magnetic Field resulting from a single loop of wire.
(c) Magnetic Field produced by a solenoid. Note that the field is fairly uniform at point P1 but not at P2.
(d) Intensifying effect of a ferromagnetic material placed inside a solenoid.

of **i** amperes will generate a magnetic field, **H**, at a distance **r** meters from the axis of the wire:

$$H = \frac{i}{2\pi r} \qquad [1.10]$$

If this wire is curved into a circular loop of radius **R** meters, the field at the center of this loop, normal to the plane of the loop is:

$$H = \frac{i}{2R} \qquad [1.11]$$

Figure 1.2b shows the new pattern of the magnetic lines of force for the looped wire. It is important to note that electrostatic lines of force start at a charged particle and end at a charged particle; such is not the case for magnetic fields. Magnetic lines of force neither start nor end, but form a continuous closed loop, as illustrated in the sketches of Figure 1.2. Also note that the magnetic lines of force are more concentrated within the loop than outside and that these lines are no longer circular (Figures 1.2b and c). This is similar to the early experiments of Ampere, in which the ends have field lines similar to a bar magnet; i.e., he formed a **magnetic dipole.**

If this single circular loop is extended to make a helical winding, or solenoid (Figure 1.2c), the field at its center and along its axis is given by:

$$H = \frac{N\,i}{L} \qquad [1.12]$$

Here, **N** is the number of turns in the solenoid and **L** its length in meters. It is important to realize that the magnetic field is fairly uniform at the center and is independent of the coil radius **R**, if the solenoid length is significantly larger than its radius. This is not true when the solenoid length is comparable to its radius, since the magnetic field is not uniform at the ends.

These magnetic fields described above exist in free space. They can also exist in matter, but a new term must be introduced to describe the response of different materials to the magnetizing field, **H**.

1.4 Magnetic Induction, B

When a demagnetized ferromagnetic material, defined later, is placed inside this solenoid, the magnetic lines of force become concentrated (Figure 1.2d). The added material appears to "enhance" or "magnify" the original magnetic field (H) that was due to the solenoid. Often, the original field is said to have "induced" the new, more powerful, magnetic field in the solid.

The "induced" magnetic field, **B**, inside a magnetic material, is given by:

$$B = \mu_0 (H + M) \qquad [1.13]$$

This is the preferred form of this relationship in SI units, according to Goldfarb and Fickett[3]. The SI unit of B is the tesla (gauss, in c.g.s.). Another form of this equation that is popular in Europe is:

$$B = \mu_0 H + J \qquad [1.13A]$$

where J is **polarization**. Note that $J = \mu_0 M$, so that both equations are entirely compatible, even though Equation [1.13A] is apparently not recognized under SI[3].

The term **M** is the **"intensity of magnetization"** or simply **"magnetization"** and has the units of tesla or weber/m^2. **M** is defined as the magnetic moment per unit volume of material. Note that **H** can exist anywhere, i.e., in "empty" space or in matter. However, **M** can exist only in matter. Traditionally, the magnetization **M** is described as the sum of all the singular magnetic moments in a unit volume of material.

In a vacuum, Equation [1.13] becomes $B = \mu_0 H$. The constant μ_0 is called the **permeability of free space** and has the value $4 \pi \times 10^{-7}$ henry/meter, in SI units (in c.g.s, this is unity). The **relative permeability,** μ_r, is the ratio μ/μ_0, which is dimensionless and equivalent numerically to the c.g.s. value for permeability.

The **permeability** of a material is defined by the ratio:

$$\mu = \frac{B}{H} \qquad [1.14]$$

While its **susceptibility** is defined by the ratio:

$$\kappa = \frac{M}{H} \qquad [1.15]$$

Both of these parameters describe how the magnetic material modifies the magnetic field inside the original solenoid. To six decimal places, the value of μ_r for air is unity, so that B essentially equals H in this case. For ferromagnetic and ferrimagnetic materials the value of μ_r essentially describes the amplification or multiplication of H by the inserted material, leading to a much higher value of induction, B. The value of susceptibility, κ, is the ratio of the density of the lines of magnetization to the lines of applied field. These two terms can, in fact, be used to characterize materials magnetically:

(a) **Vacuum**; $\kappa = 0$; $\mu = 4 \pi \times 10^{-7}$ henry/meter; $\mu_r = 1$.
Since $M = 0$, then $B = \mu_r H$, from Equation [1.13], above.

(b) **Diamagnetic materials**; κ is small and negative,
while μ_r is just less than unity.

(c) **Para- and antiferro-magnetic materials**; κ is small and positive,
while μ_r is slightly larger than unity.

(d) **Ferro- and ferri-magnetic materials**; κ and μ_r are
large and positive as well as being functions of H.

Much in the way that a transistor or diode can amplify electrical current, ferromagnetic and ferrimagnetic materials can "amplify" the intensity of the original magnetic field. These two types of magnetic materials are the most important industrially; certainly with respect to powder metallurgy and metal injection molding.

Ferromagnetic components can be made by powder metallurgy and are the vast majority of parts that are referred to in discussions of this subject. All of the discussions on high performance soft magnetic applications refer to the ferromagnetic class of materials.

One method of evaluating the response of a material to the magnetic field, H, is with the aid of magnetic hysteresis curves (see Appendix B for a list of ASTM standards and test methods). To generate a hysteresis curve, two independent coils (or solenoids) are wound on the toroidal or ring sample. The primary coil creates the applied field **H** as given by Equation [1.12], where **L** is replaced by the mean of the inner and outer ring circumferences. The secondary coil is used to measure the "induced" field, **B**, inside the magnetic material, which is given by Equation [1.13].

The hypothetical magnetic hysteresis curve shown in Figure 1.3 defines some of the basic terms that are used to characterize magnetic materials. A small applied field, H, causes a large rise in induction or flux density, B. This schematic and the proceeding hysteresis curves are plotted with B in tesla (kilogauss) and H in ampere/meter (Oersteds). For the ferrous-based materials, B is much larger than H so that the B axis has to be condensed for ease of representation.

The **initial permeability**, μ_i, is usually very difficult to measure because one is trying to determine the initial slope of the curve, and that is dependent on the sensitivity of the test apparatus and the scales used to

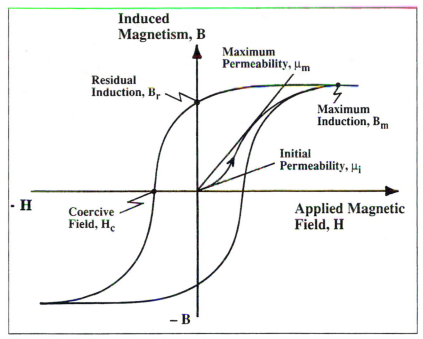

Figure 1.3: Idealized magnetic hysteresis curve.

make the graphical plot. Therefore, it is more common to refer to **maximum permeability,** or μ_m, which is the slope of the line from the origin to the tangent at the knee of the B vs. H curve.

As H is increased, the material responds with less and less "amplification" until the point B_s, **saturation induction,** is reached. After B_s has been attained and the value of H (i.e., the current in the primary field) is reduced, it is found that the initial B vs. H curve is not retraced; hence the term hysteresis. At zero applied field, there is now some retained or **residual induction,** B_r. To reduce this induced magnetism back to zero, an applied field of - H_c, or **coercivity,** is necessary. Further changes to more negative values of H lead to saturation conditions identical to B_s. As the total cycle is traversed, an entirely symmetrical curve is produced.

There is often confusion over the terms **saturation induction, B_s** and **maximum induction, B_m.** It should be understood that the **maximum induction** is simply that value of B for a **given value of applied field.** Since this is likely to be a minor loop, it is recommended that B_m should be referenced by the particular applied field chosen for the test. In a similar manner, **coercivity** and **coercive force** are the correct terms for the intersections with the H axis for the major and minor loops, respectively.

Regardless of these definitions, the terms are used interchangeably. Since H_c is an intersection of the H axis, it also appears appropriate to refer to this as coercive field as against a coercive force. In most practical devices, saturation induction is never approached so that applied fields of only 1194 or 1990 A/m (15 or 25 Oersteds) are customarily used in the P/M industry. The value of the magnetic parameters measured at these applied fields are usually sufficient to differentiate between the responses of materials under review. Of course the value of the applied field should be clearly stated, especially in the case of H_c, B_m, and B_r, since they are so sensitive to this test parameter.

In order to explain the effect of the inserted magnetic material and, therefore, the observed shape of the hysteresis curves, the concept of magnetic domains is now introduced.

1.5 Magnetic Domains

A clearer understanding of ferromagnetic materials began with two remarkable concepts introduced by the French scientist Pierre Weiss[4,5]. In his "molecular field theory" he hypothesized that:

(a) Ferromagnetic materials are "spontaneously magnetized" or "self-saturating" <u>even in the absence of an external magnetic field</u>, and

(b) These materials are divided into very small regions or magnetic "domains" which are separated by domain walls.

In the fully demagnetized state, each "domain" is magnetized to saturation by the internal-molecular-magnetic field, but the orientation of each magnetic domain is such that the net magnetization is zero. Many magnetic domains can exist in a single grain or a powder particle. Figure 1.4 is a schematic showing magnetic domains and the intervening boundary or domain wall.

The direction of magnetization in a given domain is determined by the crystal structure of the material. Crystal anisotropy leads to **magnetic anisotropy** in that certain crystal directions are easier to magnetize than others. In a given domain, the direction of saturation magnetization is one such **easy direction** of magnetization. Near room temperature, iron exists as the ferritic phase in which the crystallographic structure can be described as b.c.c. (body-centered-cubic). This is a cube, with one iron atom at each of the corners and one in the center of the cube. For b.c.c. iron, the cube edge direction, [100], is the most easily magnetized direction.

The **domain or Bloch walls** have a small but finite thickness, between

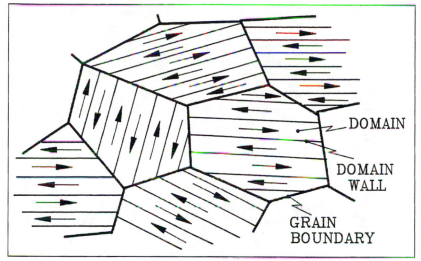

Figure 1.4: Magnetic domains inside a polycrystal - schematic.

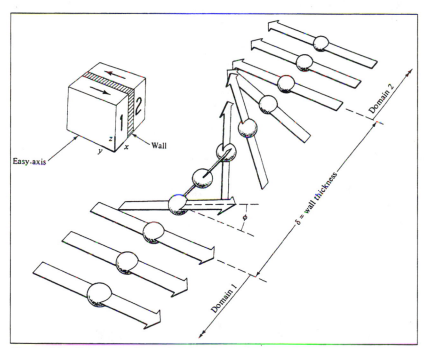

Figure 1.5: Schematic, illustrating the change in magnetic moment across a 180°
magnetic domain or Bloch wall[1].

about 0.03 to 0.2 μm. For the specific case of a 180° wall, the direction of magnetization is exactly opposite for the two domains either side of this interface (Figure 1.5). The direction of magnetization in the domain wall changes smoothly from one side to the other.

When an applied magnetic field is imposed, the first tendency is for the domain walls to move in such a manner that those domains aligned closest to the direction of the applied field grow at the expense of the opposing ones. As the strength of the external, applied, field is increased, the remaining domains are forced to rotate such that their direction of magnetization is aligned with the applied field.

In general, domain wall motion is the predominant phenomenon up to the knee of the B vs. H curve (Figure 1.6) for a polycrystalline material. Above this point, magnetic domain rotation is needed to cause a small change in B (or, more correctly, M). The schematic in Figure 1.6 should be regarded as a guide only, since some degree of domain rotation or wall motion can occur above or below this line. It should be understood that no physical movement of atoms or change in crystal structure takes place. It is the direction of the internal magnetic field (at a more fundamental level, a consequence of the nature of the electron spins and orbits) that is being modified. Since the domains are being forced to magnetize along "harder"

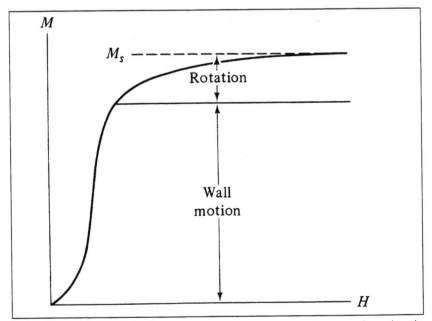

Figure 1.6: Schematic, illustrating the regions of the hysteresis curve where domain wall motion predominates over domain rotation[1].

14

crystallographic directions, the material resists this change. Above the knee of the hysteresis curve, the rotation of the miniature or elementary magnetic moments becomes increasingly more important. Eventually, the stronger external field aligns all the internal magnetic moments, and the material is said to be at saturation induction.

1.6 Energy Losses in Alternating Fields

The most uniquely important quality of soft magnetic materials is their ability to respond to very small external magnetic fields. By definition, the better soft magnetic materials have the lowest coercive fields and highest permeability. Alternating fields introduce the additional factor of time, usually in terms of the cycle frequency. The dynamic response of the magnetic material is influenced by a number of factors including crystal structure, internal defects, electrical resistivity, dimensions, test frequency, etc.

The observed behavior is related to the ease or difficulty with which domain walls are permitted to move. The motion of domain walls may be hindered by the presence of imperfections and impurities in the material. This includes the presence of inclusions such as oxides, nitrides, carbides, sulfides, etc. In addition, vacancies, pores, deformation damage (dislocation/residual strains), and any nonmetallic species will affect domain wall motion. As the applied field is varied slowly and smoothly, the motion of the domain walls is observed to be erratic and jerky. This erratic motion of the domain walls from the imperfections is called the **Barkhausen effect.** The domain walls appear to "jump" from one position to another.

The dimensional change in a sample when a magnetic field is applied to it is called **magnetostriction.** This change in length is expressed as the normalized strain ($[l_f - l_i] / l_i$) and is very small; about 10^{-5} for the common magnetic materials. At the magnetic domain level, the value of the magnetostriction is dependent on the direction of magnetization. Note that for b.c.c. iron, the [100] cube edge is the "easy" direction of magnetization. As the domain is spontaneously magnetized to saturation along such a direction, the crystal lengthens in that direction; i.e., iron has a positive magnetostriction value. These dimensional changes introduce stresses in the material surrounding the domain. Such internal stresses and strains will always be present in any magnetic material below the Curie point, no matter how slowly it may be cooled. The **Curie point** is the temperature above which a given material loses its magnetic properties.

These phenomena determine the shape of the hysteresis curve, which in turn is responsible for one portion of the energy losses. The work done in magnetizing and cycling a sample is simply the area inside the loop of its hysteresis curve (Figure 1.3). Mathematically, this is the integral of H from the extreme tips of the loop. This is called **hysteresis loss,** and clearly increases as the tips of the hysteresis curve ($+B_m$ and $-B_m$, for minor loops) become larger.

Another source of energy loss is that due to **eddy currents.** The changing applied magnetic field induces an electromotive force (e.m.f.) according to **Faraday's Law.** The direction of this e.m.f. and the associated current is, according to **Lenz's Law,** opposite to that of the external magnetic field.

The sum of these energy losses is called **core loss** (E_T):

$$E_T = E_H + E_E \qquad\qquad [1.16]$$

where E_H and E_E are the hysteresis and eddy current losses, respectively.

The hysteresis loss is dependent on the inherent crystalline properties of the material and its structural state. As indicated earlier, the width of the hysteresis curve ($2 \times H_c$) will be affected by those structural defects that inhibit magnetic domain wall motion. A larger area will be swept out as the applied magnetic field and, more correctly, the induced flux density is increased.

The magnitude of the induced eddy currents is inversely proportional to the electrical resistivity of the material. Therefore, E_E is also related to resistivity in the same way. While metallic systems show a significant eddy current effect, ferrites (which are insulators) show no such effect at low frequencies (less than 1 MHz).

As the eddy currents flow inside the sample, they also generate an associated magnetic field. This internal field is in opposition to the external magnetic field, and tends to shield the internal sections of the sample. This so called "skin effect" means that in alternating fields, only the surface of the sample interacts with the applied magnetic field and the rest of the material is essentially inactive. The degree of this shielding increases with increasing frequency.

For a strip (assuming constant permeability), the eddy current energy loss is related to these factors by the well-known Steinmetz equation[2]:

$$E_E = \frac{Kf^2B^2t^2}{\rho} \qquad\qquad [1.17]$$

where K is a proportionality constant,
          f is frequency in hertz,
          B is flux density in tesla,

16

t   is the sample thickness in meters, and
ρ   is the resistivity in ohm·meters.

In practical terms, only the electrical resistivity and the sample thickness can be modified to minimize eddy current loss. The most well-known example is power transmission transformers where silicon steels with high electrical resistivity are used in the form of thin laminated sheets. Each sheet is separated by an electrically insulating layer.

For powder metallurgy, the thickness is not a parameter that can be used as a variable to control eddy current losses. Thin strips made by P/M are very fragile in the green state. Therefore, the possibility of introducing microcracks or causing total fracture of the pressed product is high. In any case, the cost of making strips by P/M is not competitive with wrought manufacturing techniques. Electrical resistivity is therefore the only control variable that can be used in P/M to alter eddy current losses.

Beyond the core loss discussed above, an additional energy drain is the **copper loss.** This is simply the power expended in the windings; its value is I^2R, where I is the current and R is the resistance. Indirectly, the magnetic performance of the core material (specifically coercive field and permeability) affects the copper loss since a magnetically softer material will require less power to achieve the same flux density. All of these energy losses are exhibited in the form of heat loss.

1.7 Internal Demagnetizing Fields

One of the most difficult effects to account for in magnetic testing is that of demagnetizing fields at the open ends of test samples. This is not the case for ring or toroidal shaped-samples. There are no ends to the ring, so that the induced flux lines are continuous and completely inside the material. This is termed a **closed magnetic circuit;** an **open magnetic circuit** is one in which the flux lines are partly in air and partly in the magnetic material.

Consider a bar that was magnetized by an external field, H, which was then removed[1]. In the absence of any external influences, magnetic field lines will radiate from the north pole to the south pole (Figure 1.7a). We are familiar with the field lines on the outside of the magnet. However, the field lines on the inside of the magnet also run from the north to the south pole, but are opposite in direction and, therefore, act to **demagnetize** the magnet. Since H_d acts in the opposite direction to M, the induced magnetism inside the magnet is

$$B = M + \mu_0 (H - H_d) \qquad\qquad [1.18]$$

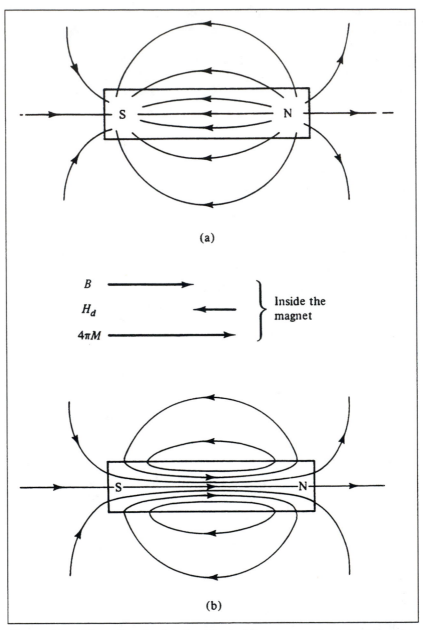

Figure 1.7: Schematic, showing the effect of the demagnetizing field in a Bar magnet[1].

    a) Magnetizing (H) and demagnetizing fields (H_d).

    b) Magnetizing (H) and induced fields ($B=4\pi M-H_d$).

As shown in Figure 1.7b, the flux lines for B are continuous and track from a south to a north pole. The **demagnetizing field,** H_d, is strongest near the poles and results in a lesser flux density at these sections of a bar magnet[1].

The demagnetizing field, H_d, is directly proportional to the magnetization, M, as follows:

$$H_d = N_d \times M \qquad\qquad [1.19]$$

The proportionality constant, N_d, is called the **demagnetizing factor or coefficient.** This is essentially dependent on the shape or geometry of the part. An accurate value for N_d can only be calculated for the special case of

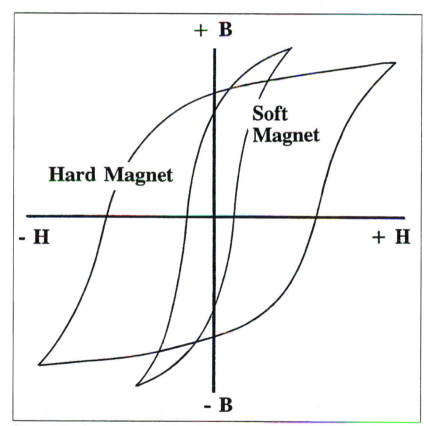

Figure 1.8: Schematic, illustrating the difference in properties of hard and soft magnetic materials.

the ellipsoid. For other shapes, the value is estimated or measured experimentally with a search coil (see section 1.10.1).

1.8 Soft or Temporary Magnets

In practical terms, the best **soft or temporary magnetic materials** have a narrow hysteresis loop (Figure 1.8). With regard to the basic measurable properties, this means that H_c should be as small as possible, while μ_m and B_m should be as large as possible. Another way of looking at this is that the goal is to get the maximum output (B) with the smallest energy input (H). In general, B_r should also be minimized, unless a reverse current is going to be applied during the application cycle. From an energy utilization standpoint, the area inside the loop is related to the work done in traversing the cycle. This is released as heat.

In addition, eddy currents are set up inside the magnetic material in response to the applied field, as discussed in section 1.5, above. These eddy currents and the heat buildup due to them is inversely related to the electrical resistivity of the material. Therefore, the electrical resistivity of the material should be as large as possible.

1.9 Hard or Permanent Magnets

When fabricating **hard or permanent magnetic materials,** the objectives are entirely opposite with regard to coercive field. As shown in Figure 1.8, H_c should be as large as possible. The goal is to minimize the influence of external magnetic fields; the hard magnetic material should not change its characteristics if an external field is brought near it. The following is a brief introduction to this sister field, simply so that one may appreciate the aspirations of the magnet circuit designer.

Discussions on permanent magnetic materials focus on the second quadrant of the hysteresis curve (Figure 1.9), often referred to as the demagnetizing curve. For these materials, the value of applied field H becomes so large that there is a significant difference between the **normal** and the **intrinsic** properties. Figure 1.9 shows both a normal and an intrinsic demagnetization curve. The intrinsic induction is simply the normal induction, M, plus the applied field, H.

To obtain the normal induction, H is subtracted from B_i. In the soft magnetic case the value of H is sufficiently small to be ignored. In hard magnets, such is not the case. Therefore, the two curves shown are traditionally used; the normal curve for determining the flux density in the air gaps, while the intrinsic curve is used for determining the effects on the magnet itself as a result of any external demagnetizing conditions. The

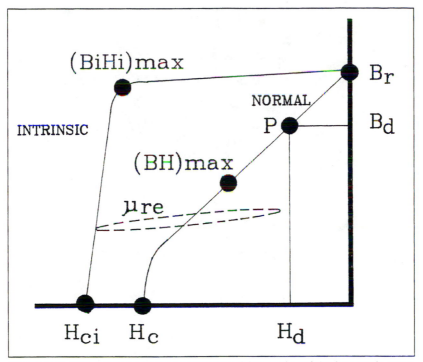

Figure 1.9: The second quadrant or demagnetizing portion of the hysteresis curve for a permanent magnet - schematic.

higher the value of the coercive field (and therefore the higher the applied field to attain this), the more divergent are the normal and intrinsic curves.

Similar to Figure 1.3, B_r is the residual induction. H_{ci} and H_c are the intrinsic and normal coercive fields, respectively. The **maximum energy product** $(B \times M)_{MAX}$ and the **recoil permeability** (μ_{re}) are also additional parameters defined in the second quadrant (Figure 1.9).

From a performance perspective, a high B_r value means that the magnet will have a high strength and a high H_{ci} means that it will be less easily affected by demagnetizing conditions. The maximum energy product is used as a **figure of merit** to make a quick performance comparison between the available magnetic materials. The recoil permeability is used to determine the behavior of the magnet after a demagnetizing field has been applied. This parameter shows the track that B and H will follow on the hysteresis curve.

The task of the magnetic circuit designer is to first determine where on the demagnetization curve the part will operate. Then the design of the

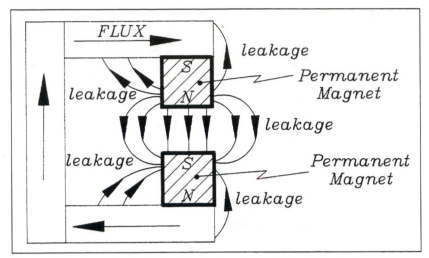

Figure 1.10: Magnetic circuit showing the importance of flux leakage - schematic.

magnetic circuit is modified so that, ideally, the operating point of the magnet will coincide with the maximum energy factor (BH) for optimum performance. Finally, the components are modified for shape, size, etc., in order to make the assembly as economical as possible.

The attainment of these optimum conditions is not easy, however, since both the **leakage** and **reluctance** factors (see below) can vary dramatically. Often, designers base their magnetic circuits on earlier experiences, published work or finite element analysis software programs.

A simple magnetic circuit is shown in Figure 1.10, consisting of permanent magnets which are attached to the soft magnetic components that transmit the flux and act as a return path. Ideally, the magnetic flux should form a continuous loop. In reality, the flux leaks outside the magnetic circuit and also spreads out at the air gap. This is accounted for by the **leakage factor, F,** which can range anywhere from 1 to 20. In addition, the **reluctance factor, f** (which usually ranges from 1 to 1.3), relates to the field decreases at sites away from the air gap. This includes decreases due to joints and the return path material being near saturation and not being able to carry the required field density.

The operating load line at any point P on the normal curve of Figure 1.9, is given by the following:

$$P = \frac{B_d}{H_d} = \frac{l_m}{A_m} \cdot \frac{A_g}{l_g} \cdot \frac{F}{f} \qquad [1.20]$$

where A_m is the permanent magnet's cross sectional area,

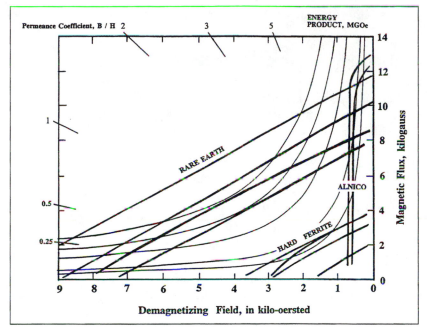

Figure 1.11: Properties of some popular permanent magnetic materials.

l_m its length,
A_g is the cross-sectional area of the air gap, and
l_g is its length.

From this equation it can be seen that an increase in F is comparable to an increase in the air gap area (A_g), while an increase in f is the same as increasing its length (l_g). Clearly, this analogue assumes that the magnet dimensions are kept constant. These are the parameters that a designer alters in order to optimize the energy product (BH). More detailed information is provided in a number of excellent publications[6,7,8].

Different materials exhibit a variety of demagnetization curves, as illustrated in Figure 1.11[11]. Stadelmaier and Henig[10] reviewed the latest trends in permanent magnet materials and concluded that research on $Fe_{12}Nd_2B$ -based alloys continues, but new rare-earth alloys based on $Fe_{27}Nd_2N$ appear to hold promise of higher Curie temperatures.

In summary, the principle aim in producing hard magnets is to **__minimize__** grain size (more specifically, magnetic domain size) and firmly "pin" these in place, after appropriate orientation in a strong external magnetic field. This "pinning" is aided by the presence of particles or precipitates. Furthermore, deformation in the form of dislocations at the microcrystalline level can also be effective in the pinning of domain walls.

1.10 Measurement of Magnetism

In this section, certain aspects of magnetic testing are discussed in a very brief and introductory manner. In principle, almost any shape or size of part can be checked for magnetic characteristics. However, practical limitations dictate that only a few test procedures provide useful and accurate data for component design. A few examples are illustrated here; Appendix B lists some selected test methods and material specifications that have been agreed-to by ASTM and the user-community. The reader is encouraged to consult these documents for detailed information on any specific subject.

1.10.1 *Detection of Magnetic Fields*

A magnetic field can be detected by a **search coil** placed perpendicular to the flux lines. The magnetic field generates an electromotive force, e.m.f., by Faraday's Law, according to the rate of change of flux:

$$e = -N \frac{d\Phi}{dt} \qquad\qquad [1.21]$$

where N is the number of turns on the coil. The coil is moved quickly from the vicinity of the magnetic field or rotated. The SI units of magnetic flux, Φ, are webers. The search coil is connected to a **ballistic galvanometer** or a **fluxmeter,** which detects the current associated with the e.m.f.

A **Hall probe,** also called a **search probe,** is based on the principle that a magnetic field alters the flow of current through a metal or semiconductor. The latter are used most often in these probes because they exhibit a much larger Hall effect. The probe is connected to a source of constant current and an electronic circuit for measurement and amplification of the associated e.m.f. A Hall probe needs to be calibrated with a known magnetic field. An advantage of this probe is that it can be very small and that the magnetic field can be static, i.e., the flux does not need to be fluctuating as in the case of the search coil.

A **vibrating sample magnetometer** is based on detecting the change in flux of a coil in the vicinity of a moving sample. The sample is positioned between the poles of a magnet and the sample vibrated normal to its flux lines. Detecting coils are positioned near the test sample. A similar set of coils detects the flux changes in a hard magnet reference sample. Both the reference and the test sample are vibrated at the same frequency and the e.m.f. values from the two coils compared.

1.10.2 *Hysteresis Curves*

Ring samples can be tested without the need to be concerned about demagnetizing effects. From this perspective, the measurement of magnetic properties from hysteresis graphs using ring samples is the preferred form of testing. Appendix B lists several standard test methods that have been developed by ASTM, including the hysteresis graph procedure. Briefly, the procedure involves the fabrication of a ring or toroid of known dimensions, on which are wound two separate windings, as discussed in Section 3.1. The applied magnetic field is given by Equation [1.12], above. The number of turns on the primary windings is substituted for "N", and "L" is replaced by the average of the inside and outside circumferences of the ring. The applied field can be calculated, using this equation and a given value for current. The induced magnetic field is measured with a ballistic galvanometer or fluxmeter.

1.10.3 *Epstein Frame*

An **Epstein frame** simulates a transformer in that a square frame is made from strips of the test material. The total mass of the strips must be known as this is used in the calculations. Primary and secondary coils are wound on each of the four sections and the power losses checked in the no-load condition (secondary circuit open). The power loss is measured at a given frequency and flux density using a wattmeter (electrodynamometer or electronic digital type). The core loss is reported in watts/kg or watts/lb, at a specified frequency and flux density. Appendix B lists the ASTM standard test methods that pertain to the Epstein frame.

1.11 Customary Units

In many parts of the world, the International System of Units (SI) is not fully in place, and the "customary" c.g.s. units may still be popular with those working on magnetism. For this reason, the following magnetic formulae are shown with c.g.s. units. These are the "customary" equivalents of the formulae that were used in the text above.

Wire, $\qquad H = \dfrac{2i}{10r}$ $\qquad\qquad$ Oersted $\qquad$ [1.10C]

Loop, $\qquad H = \dfrac{2\pi i}{10R}$ $\qquad\qquad$ Oersted $\qquad$ [1.11C]

Solenoid, $H = \dfrac{4\pi Ni}{10L}$ Oersted   [1.12C]

Induction, $B = H + 4\pi M$ Gauss   [1.13C]

The **permeability** of a material is defined by the ratio:

$$\mu = \frac{B}{H} \qquad [1.14C]$$

While its **susceptibility** is defined by the ratio:

$$\kappa = \frac{M}{H} \qquad [1.15C]$$

Both of these parameters describe how the magnetic material modifies the magnetic field inside the original solenoid. Since the value of μ for air is unity, B essentially equals H, in this case. For ferro- and ferri-magnetic materials the value of μ essentially describes the amplification of H by the inserted material, leading to a much higher value of induction, B. The value of susceptibility, κ, describes how responsive or sensitive the material is to an applied field H. These two terms can be used to characterize materials:

  (a) **Vacuum**; $\kappa = 0$, $\mu = 1$
     Since M = 0, then H = B, from Equation [1.13C], above.

  (b) **Diamagnetic materials**; κ is small and negative, while μ is just less than unity.

  (c) **Para- and antiferro-magnetic** materials; κ is small and positive, while μ is slightly larger than unity.

  (d) **Ferro- and ferri-magnetic materials**; κ and μ are large and positive as well as being functions of H.

$$B = (H - H_d) + 4\pi M \qquad [1.18C]$$

The demagnetizing field, H_d, is directly proportional to the magnetization, M, as follows:

$$H_d = N_d \times M \qquad [1.19C]$$

The proportionality constant, N_d, is called the **demagnetizing factor or coefficient.** In the case of a cylindrical part, the demagnetizing factor along its length tends to zero as the length/diameter ratio approaches infinity. Normal to this axis, the value of N_d is $4\pi/3$. As an example, for an l/d ratio of 20, the value of N_d is 0.25. If B = 10,000 Gauss, then M = 10,000 / 4π = 796 emu/cm^3. Rounding this to 800, the value of H_d is $N_d \times M = 0.25 \times 800 = 200$ Oersteds. This is a significantly large value, considering that the intrinsic coercive field for most soft magnetic materials is less than 100 Oersteds. The point to emphasize is that the demagnetizing field may play a significant role in magnetic circuit design and steps must be taken to correct for it.

REFERENCES

1. B. D. Cullity, *Introduction to Magnetic Materials*, Addison-Wesley, Reading, MA, 1972.
2. R. M. Bozorth, *Ferromagnetism*, Van Nostrand, New York, NY, 1951.
3. R. B. Goldfarb and F. R. Fickett, NBS Special Publication 696, U.S. Government Printing Office, March 1985.
4. Pierre Weiss, "La Variation du Ferromagetisme avec la Temperature", *Compt-Rend*, 1906, Vol. 143, pp. 1136-1139.
5. Pierre Weiss, "L' Hypothese du Champ Moleculaire et de la Propriete Ferromagnetique", *J. de Physique*, 1907, Vol. 6, pp. 661-690.
6. M. A. Bohlmann, "Rare Earth Permanent Magnet Materials for Motion Devices", *Power Conversion and Intelligent Motion*, 1987, Vol. 13, pp. 48-56.
7. MMPA Standard No. 0100-87: "Standard Specifications for Permanent Magnet Materials", 1987, Magnetic Materials Producers Association, Evanston, IL, 28 pages.
8. "Guide to ... Permanent Magnet Materials", 1988, Magnetic Materials Producers Association, Evanston, IL, 32 pages.
9. D. Hadfield (Ed.), *Permanent Magnets and Magnetism*, John Wiley, New York, NY, 1962.
10. H. H. Staidelmaier and E. Th. Henig, "Permanent Magnet Materials-Developments during the last 18 Months", *J. of Metals*, 1991, Vol. 43, No. 2, pp. 32-33.

Chapter

II.

FUNDAMENTALS OF POWDER METALLURGY AND METAL INJECTION MOLDING

Powder metallurgy and metal injection molding are used to fabricate large quantities of soft magnetic components. These forming technologies are discussed in this chapter, with particular emphasis on those process parameters that may alter the soft magnetic properties of the sintered product.

2.1 Introduction

The Powder Metallurgy (P/M) process is distinctly different from the traditional wrought metallurgy processes because the former relies primarily on solid state diffusion, as against the fusion of molten metal species. In wrought metallurgy, the raw materials are generally melted at high temperatures then cast into the ultimate shape or into an ingot for subsequent rolling. Other metalworking operations may be needed.

In conventional P/M, the melting point of the major constituents is not exceeded during the sintering step. In some cases, minor constituents may melt and form liquid phases for a short period of time. However, the shape of the initial "green" compact is not lost. In its simplest form, this fabrication technology can be considered to have four basic steps:

1) powder production,
2) compaction,
3) sintering, and
4) secondary operations.

This gross simplification belies the science and the complexity of a difficult and unique processing technology. In many cases, practitioners of this technology consider it as an art form and execute it to different degrees of competency. This leads to substantial differences in end-product properties from different suppliers, even though the input materials may be identical. Because of the extreme versatility of the process, a P/M parts fabricator can truly make a mundane article or a genuine "masterpiece".

A related but different forming technique is that of Metal Injection Molding (MIM), which can be considered to be a subset of the more general, and recently-introduced, term of "Powder Injection Molding" (PIM)[1]. This emerging parts fabrication technology gained considerable prominence in the decade of the eighties, because of the promise it offers to produce shapes and materials that would otherwise be very difficult to fabricate.

Both the conventional powder metallurgy and the metal injection molding processes are excellent techniques for producing high performance components for soft magnetic applications, beyond the traditional wrought metallurgical routes. Therefore, the following sections focus on these two technologies to promote an understanding of the science behind these fields and to gain an appreciation of those parameters that influence the final product.

2.2 Powder Metallurgy Overview

Figure 2.1 depicts the powder metallurgy (P/M) process in schematic form. Numerous research papers and classical text books, such as Jones[2], Lenel[3], and German[4] provide excellent in-depth descriptions of the metallurgical, chemical, and physical phenomena occurring at each stage of this process. The following brief descriptions are based primarily on these texts.

(a) *Powder Production*. Several methods[3,5,6,7] are used for the commercial production of metal powders for magnetic applications. These include atomization of a molten metal stream, oxide reduction, chemical decomposition, electrodeposition, milling of embrittled materials, etc. By far, the greatest tonnage of metal powder is manufactured by the atomization[6,7] of a molten metal stream. This may be accomplished by the use of jets of water or an inert gas. Oil atomization has been used in to order to minimize oxidation of the powders. However, the pyrolysis of oil leads to residue that results in excessive carbon levels. The process has been discontinued for low alloy steels, because of economic reasons.

Water atomized powder contains excessive levels of oxides

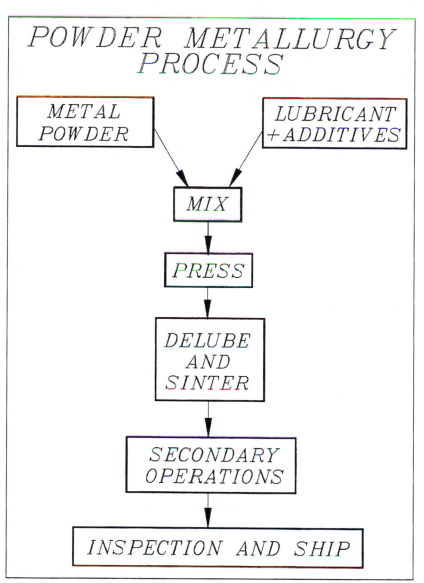

Figure 2.1: Flow diagram for conventional Powder Metallurgy (P/M) process - schematic.

which must be reduced. This is followed by an annealing step. Intermediate milling and screening steps may be necessary before re-blending and packing. Re-blending of powders of different sizes enables the optimum distribution to be achieved for compressibility and flow rate purposes.

Elemental powders or pre-alloyed powders (where each powder particle has the gross chemistry of the alloy) are then mixed to achieve the composition of the desired final product. Often this is done at the powder producer's facility for large batches; a lubricant is also mixed in at this time.

(b) *Compaction*. The traditional method of compaction in rigid tooling at room temperature is the most prevalent form of consolidation for iron-based materials. A lubricant is necessary in the powder blend, to enable the pressed "green part" to be ejected from the die. Spraying the inside of the die-cavity with lubricant, to permit "die-wall lubrication", can be done, but is not practiced frequently.

Compaction or "molding" pressures are generally in the 414 to 828 MPa (30 to 60 tons per square inch) range, for ferrous powders. Uniaxial compaction with the tool punches moving vertically has the limitation that no features involving undercuts in the horizontal direction are permitted. The sides of the dies and, therefore, the green part are generally straight and parallel, so that the tooling members and the part may be removed. Some degree of taper is permitted, provided the part can be ejected. Generally, tooling members do not move horizontally in conventional P/M presses. All features in the top and bottom surfaces of the green part are formed in the punch detail, die detail ("shelf-die"), or by the use of separate punches. Such a part would be upgraded from a "single-level part" to a "multi-level part".

Powder Forging (P/F) can be performed by pressing a heated pre-form with the tools either at room temperature or at a slightly elevated temperature. Non-rigid methods include cold isostatic compaction or pressing (CIP) in a rubber mold and hot isostatic pressing (HIP) with the powder encapsulated in a glass or a metal "can". The former uses water or oil as the pressure transfer fluid, while inert gases are used at the higher temperatures for HIP.

(c) *Sintering*. While some limited mechanical interlinking occurs during the compaction step, the real development of interparticle bonds takes place at the sintering operation[3,4]. The first goal is to remove the lubricant by thermal degradation and oxidation. Improper lubricant removal can lead to additional carbon in the product. Even small amounts of carbon residue can degrade soft magnetic properties, but may not affect mechanical properties.

The bonding of particles is assisted by using an inert atmosphere and a reducing gas such as hydrogen or carbon monoxide to remove surface oxides. Dissociated ammonia, cracked methane, nitrogen and hydrogen are utilized most often for this purpose. Furthermore, the degree of atomic diffusion is related to the sintering temperature and time. Of course a higher level of diffusion promotes interparticle bonding, spheroidization of pores, alloy homogenization, grain growth, etc. The use of atmospheres containing carbon should be discouraged, since the absorption of this element can cause significant degradation of soft magnetic properties. Vacuum sintering is also performed, often with a backfill of nitrogen, argon, hydrogen or a combination of these gases. Furthermore, vacuum furnaces can easily incorporate gas quenching systems for enhanced cooling rates.

(d) *Repressing, sizing and coining.* In order to control the product dimensions, a second repressing operation may be necessary to meet specified print dimensions. A full-form repressing will achieve this objective as well as causing additional densification. A surface detail of shallow depth can also be pressed in at this time. In some cases, partial refinement of selected dimensions can be achieved. For example, external features can be maintained to tight tolerances by pushing the part through a die. In the case of holes, the part can be pushed over a mandrel or a precision-ground ball can be pushed through the internal diameter. The latter operation is referred to as "ball-sizing".

(e) *Secondary Operations.* Many additional post-sinter operations are often performed on P/M parts. These include such diverse operations as machining, tapping, drilling, grinding, plating, coating, resin impregnation, heat treatment, etc. Infiltration of the pores with low melting point metals, such as copper, can be performed in the initial sinter or a separate thermal operation. Such an infiltration step is usually done to improve structural performance, but not for soft magnetic performance.

(f) *Inspection, Packing and Shipping.* After the final opportunity to check the part for the agreed-to criteria, the resultant product is packaged and put into stock or shipped to the customer.

2.3 Discussion of Significant Factors in P/M

While it is true that each step of the process is important, the two most significant ones controllable by the parts fabricator are compaction and sintering. Assuming that the raw ingredients are pure, and mixing is done in a reasonably controlled manner, the compaction pressure, sintering temperature, sintering time, and sintering atmosphere will dictate the density and microstructural integrity of the end-product.

The apparent density or the "filling capability" of loosely-packed ferrous powders is low (generally, about 3000 kg/m³). During compaction, the apparent density increases with compaction pressure, as the part shape is being formed. Figure 2.2 shows typical compressibility curves for some selected metal powders. These curves are dependent on the initial hardness of the raw powders as well their work-hardening response. Furthermore, the powder characteristics such as particle shape, size and distribution, interparticle friction, etc., exert a strong influence on the form of the curves shown in Figure 2.2.

The density of the green compact may not be uniform throughout, but can vary according to location with respect to the tooling and the part shape. To a first approximation, the sectional density will be highest near a tooling member. The interparticle friction and mechanical interlocking

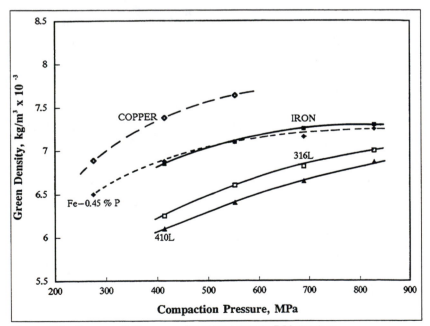

Figure 2.2: Compressibility curves for some common P/M powders.

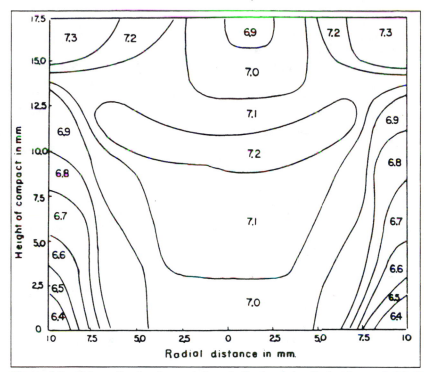

Figure 2.3: Schematic showing variations in the local density inside a cylindrical sample of nickel pressed from the top[3].

prevent the powder from behaving in a hydrostatic manner. Figure 2.3 illustrates the density variations that are observed for a cylindrical nickel powder compact, which was pressed from the top only. While some minor changes will occur during sintering, these variations in density will persist through to the final product. From a performance standpoint, it should be recalled that the flux density will be limited by the lowest density section in the flux path of the soft magnetic device.

The powder compact has some green strength that is attributable to the mechanical interlocking of particles and some limited cold welding. The specific value of this green strength is again traceable back to the particle characteristics such as shape and size, as well as the response of the material to strain. Higher green strengths are achieved by using ductile materials and highly non-spherical, irregular-shaped, powders. The importance of green strength should not be overlooked since this property dictates how well a green product will fare through the material handling systems of a normal industrial environment. A particularly low value of green strength offers enhanced probability for the initiation of microcracks or complete rupture of whole sections of parts.

Modelling of the real-world behavior of powders necessitates the simplification of a very complex system. Such is certainly the case for sintering. The classical theory of sintering[3,4] begins with discussions on the behavior of mono-sized particles of a single element, in order to eliminate the complications due to compositional variations. A further simplification is that the particles are assumed to be undeformed spheres. Even with these departures from reality, important predictions can be made about the response of metal powders to thermal processing.

Material transport occurs by a number of well-recognized methods[3,4], as shown in Figure 2.4:

- surface diffusion on the spheres
- surface diffusion along the grain boundaries
- volume or bulk diffusion in the body of the sphere
- vaporization and condensation in the pores

Ashby[8] proposed the concept of using sintering diagrams to provide guidance as to where certain material transport mechanisms were prevalent. Figure 2.5 shows such a sintering map for copper spheres[3]. Two spherical particles next to each other will initially have a point contact. These will eventually develop into a neck or a circular interface between the particles. The neck ratio (radius of the neck interface/radius of the particles) is plotted against the sintering temperature. The solid lines indicate the area at which each of the two mechanisms contribute equally to material flow. If one traverses away from the solid line into a given region, the mechanism denoted in that region of the graph becomes predominant.

For most areas on this diagram, surface diffusion and grain boundary diffusion predominate. Neither of these mechanisms will cause densification during sintering. In order to promote densification, the centers of the spheres involved (Figure 2.4) must move closer together; this can only occur by enhancing volume or bulk diffusion. As seen in Figure 2.5, volume diffusion is predominant at high temperatures and high neck ratios. High neck ratios are achieved in the green state by simply increasing the compaction pressure. However, practical compaction pressures rarely exceed 690 MPa (50 tons/in²) because of the increasing possibility of tool fracture. These influences of compaction pressure and sintering temperature are important, since the density of the final product is one of the major determinants of magnetic performance.

Thus far, discussion of the impact of transient liquid phases[3,4] has been intentionally excluded. Yet, this is one of the most important mechanisms for the sintering of alloys in the Fe-P and Fe-Si systems. The ferro-

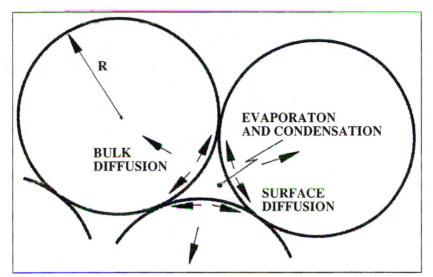

Figure 2.4: Schematic showing some of the different mass transfer mechanisms in a P/M compact.

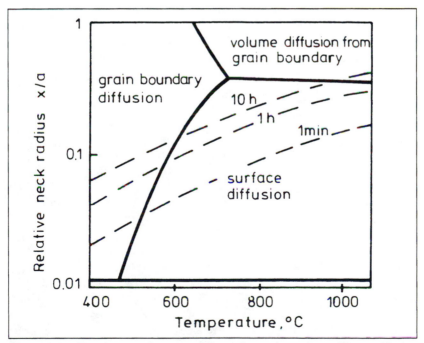

Figure 2.5: Sintering map for copper, illustrating regions where different diffusion modes predominate[3].

phosphorus, Fe_3P, that is used as the source for P in the former alloy system begins melting near 982°C (1800°F). The amount of the liquid phase initially increases with increasing temperature. As diffusion of the P into the base Fe proceeds, the melting point of the P-rich phase rises, so that eventually it too becomes a solid at the sintering temperature.

In a similar manner, the ferrosilicon masteralloys (usually containing 9 or 17% Si, by weight) exhibit a transient liquid phase above 1204°C (2200°F). The degree of shrinkage observed during sintering is directly related to the amount of the liquid phase at the sintering temperature. The presence of the transient liquid phase increases the rate of sintering because of a) the faster diffusion rate in the liquid compared to the solid, and b) the flow of material into the interstices and angled regions of porosity. In summary, the transient liquid phases enhance the sintering rate, but at the expense of dimensional stability.

A recent review[9] of the fundamentals of the sintering operation focussed on the influence of sintering atmospheres and temperatures on the properties of stainless steels and soft magnetic materials. Table 2.1 lists some of the more important phenomena that can be occurring in this operation. Note that many of these are **reversible** reactions, i.e., the specific process conditions will dictate the direction of the reactions. For example, too much moisture or oxygen in the sintering atmosphere will promote oxide formation or the decarburization of steels. The interaction of these phenomena, dictates the quality of the resultant product.

Table 2.1 Sintering Phenomena In Powder Metallurgy And Metal Injection Molding

1) Oxide reduction	7) Hydriding
2) Oxide formation	8) Homogenization
3) Oxide dissolution	9) Recrystallization
4) Carburization	10) Grain growth
5) Decarburization	11) Increase in bond area
6) Nitriding	12) Change in pore structure

The removal of the organic lubricants should be carried out at low temperatures, and at relatively high dew points. This aids their removal by oxidation, while minimizing chances of pyrolizing the organic constituents of the lubricants. Special emphasis must be placed on the importance of the thermodynamics of the reduction process for the metal oxides[9]. It can be shown from first principles that this reduction process is aided by higher temperatures and lower dew points, Figure 2.6. This figure displays the equilibrium conditions of the oxidation-reduction reaction for a variety of

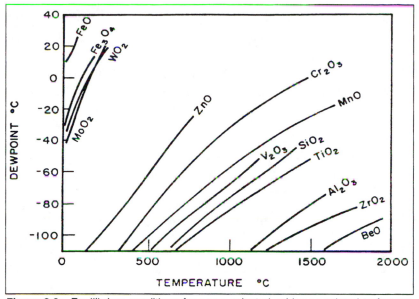

Figure 2.6: Equilibrium conditions for some selected oxides as related to furnace dew point and sintering temperature[3].

oxides. To the top and left of a given line are oxidizing conditions, while reducing conditions prevail to the bottom and right of the curve. These equilibrium curves are predicted from thermodynamic calculations and not from kinetic considerations. This implies that Figure 2.6 displays what can happen thermodynamically, but does not reveal if sufficient kinetic energy is available to allow that reaction to take place. This is especially important at the lower temperatures.

Specifically, the oxides of Cr (in stainless steels) and those of Si (in magnetic Si-Fe steels) are very stable and require low dew points and high temperatures. Just as important as maintaining a low (reducing) dew point in the high heat section is the need to keep this variable as low as possible in the cooling section of the sintering furnace. Otherwise, re-oxidation will occur at this stage of the sintering cycle.

Post-sinter operations such as repressing, machining, drilling, milling, grinding, tapping, etc., will degrade soft magnetic performance. These types of operations cause deformation damage or strain that inhibits the motion of magnetic domain walls. At the microstructural level, the dislocations that are generated by the deformation damage "pin" the domain walls in the same manner as grain boundaries. This deformation damage can be removed by simply annealing at a low temperature. Such a treatment will restore soft magnetic properties very effectively. Care must be taken so as not to introduce fresh damage in the form of thermal strains during the cooling cycle. Specific thermal profiles are recommended for certain materials (see Appendix B for a listing of some ASTM documents that may provide guidelines).

In a.c. type applications, device performance is very sensitive to surface damage because of the "skin effect". In such applications only the surface of the soft magnetic component has time to respond to the external magnetic; the bulk of the part does not participate in its function. Therefore, it is highly recommended that such a component be thermally processed to anneal out the deformation defects. Such an anneal should be performed in a reducing atmosphere (preferably hydrogen) and the part cooled slowly to avoid the re-introduction of damage by thermal stresses.

To a first approximation, B_m and resistivity are dependent on the material and its density (**"structure insensitive"**). On the other hand, the other magnetic properties (B_r, H_c and μ_m) are more sensitive to the part structure (**structure sensitive**). Specifically, these properties will be improved to the degree that the part's fabricator controls impurity levels (especially interstitials such as C, N_2, O_2 and S), deformation damage, and microstructure (grain size, precipitates, etc.).

2.4 Metal Injection Molding Overview

Metal Injection Molding (MIM) has evolved into a production process for the fabrication of complex-shaped parts, primarily as a result of the efforts of Raymond E. Wiech, Jr., in the decades of the seventies and early eighties. The binder systems and the technology utilized were regarded as proprietary until the issuance of three U.S. patents[10,11,12].

Much of this pioneer's work was, however, in the Ceramic Injection Molding (CIM) of high purity alumina parts for the Parmatech Corporation of Petaluma, California. Production components included wire guides and furnace substrates. Ceramic injection molding as a technology is not new, however, as this fabrication method was developed by Schwartzwalder[13] in 1937 for producing spark plug insulators. The fact that a competing technology ("slip casting") was chosen as the production method, is incidental. Much of the technology used in MIM is identical to CIM. One major distinction is that ceramic powders are non-conductive and therefore hold electrostatic charges. These create repulsion forces and difficulties that need to be overcome during binder system development. Two outstanding texts[14,15] were used extensively in a recent development program for CIM. The conclusion of this work[16] was that highly intricate components of high purity alumina could be made on a production basis to tolerances of 0.2% (+/- 0.002 inches/inch) with greater than a 99.9% confidence level.

40

Mangels and Trela[17] have given a review of some of the more important parameters in CIM for making high performance gas turbine engine components. Hunold, et al.[18] described the manufacture of rotors from silicon carbide and silicon nitride, offering great promise for the process to be successful in producing such high performance components.

Several companies initiated MIM programs in the late seventies and early eighties. Pease[19,20] estimated that some 30 companies were active in MIM in the USA at various levels of proficiency and capitalization. While some of these companies have not been successful and new ventures have been initiated, a few companies have continued fairly strong. A complete list and survey of company activities are provided by Pease[20].

Remington Arms Company, a business unit of DuPont, initiated its efforts in MIM and CIM in 1980 with internal R & D programs and as a consortium member of an MIM program at Battelle Laboratories in Columbus, Ohio. Multicomponent binder systems were developed and a production process for MIM parts put into full effect by 1984. Much of the technology is considered proprietary and cannot be discussed here, but the operation continues as a production parts business for both commercial needs as well as in-plant requirements. Several soft magnetic components are being produced, in shapes and compositions that conventional P/M could not handle[21].

The definitive text currently available on the subject of MIM and CIM is "Powder Injection Molding", by German[1]. While the term "Powder Injection Molding" (PIM) is perfectly appropriate in a general discussion of both topics, it is believed that practitioners of MIM and CIM need to understand that these are significantly different classes of materials. As such, the scientific basis to process such materials must take into consideration the differences in material behavior. While theoretical studies on packing efficiencies will not be affected, work on binder development and sintering most definitely will need to reflect the sensitivity of the particulate material to its surroundings. As an example, ceramic particles in the sub-micron range can be handled routinely, but metal powders may be pyrophoric so that protective atmospheres must be used. Similar considerations must be given to the sintering environment.

2.5 Discussion of Significant Factors in MIM

The metal injection molding process (Figure 2.7) begins with the intimate mixing of fine metal powders with a binder system. This composite is formed into a shape in a conventional plastic injection molding. The molded green part is processed to remove the binder system and is subsequently densified at the sintering stage. Beyond this, the normal finishing operations (machining, grinding, plating, heat treating, etc.) may

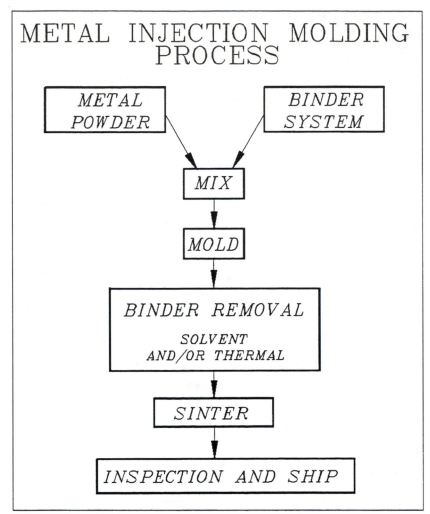

Figure 2.7: Flow Diagram for the Metal Injection Molding (MIM) process - schematic.

be utilized.

Powders for conventional P/M range from about 20 to 200 microns, whereas powders for MIM are routinely less than 20 microns. CIM powders are even smaller, preferably less than a micron in diameter. The need for fine powders in both MIM and CIM is that the high surface energy is all that provides the driving force for consolidation (during sintering). In certain material systems, transient liquid phases will assist densification. Without this possibility, the resulting product will have excessive porosity.

If a low density product is desired, coarser powders may be used. Fine powders are not needed from the perspective of aiding feedstock flow or moldability of the green part.

This is in contrast to P/M, where most of the consolidation/densification occurs during the initial, mechanical, compaction stage. Also, fine powders in P/M create difficulties with flow and by lodging in between tooling members. The result of this may be tool wear or failure, as the tools get "locked" in eject or mold positions.

Theoretically, for injection molding, the finer the powder the better, since this results in an increased sintered density. In practice, however, there are tremendous difficulties in producing and handling very fine metal powders. First, the "atomization" of liquid streams by a gas or liquid yields very low fractions of powders less than 325 mesh (less than 45 microns). Even though this is very inefficient[22], this is the main source of fine stainless steel powders for MIM; these are simply screened out from the P/M grade powders. Lawley[7] discussed the ability of high pressure water atomization (greater than 20 MPa) with special jet designs to produce fine metal powders. Furthermore, special nozzles aided by ultrasonic gas atomization at pressures near 12 MPa can also produce such fine particles. While such techniques may not be fully commercialized, they offer the promise of the ready availability of fine powders in the near future.

The major production route for fine metal powders is still chemical, which includes oxide reduction, electrolytic deposition and chemical decomposition[3,4]. The oxide reduction process involves the reduction of high purity Fe_3O_4 (magnetite) by carbon monoxide or hydrogen[4], yielding a product called "sponge iron". As the name implies, electrolytic deposition involves the deposition of the metallic ions from solution onto the cathode. The deposit is removed from the electrode and ground to the required particle size. A final anneal to remove deformation damage results in a reasonably high purity product.

Fine iron and nickel powders are made from the thermal decomposition of iron carbonyl ($Fe(CO)_5$) and nickel carbonyl ($Ni(CO)_4$), respectively. The resulting product is pure and the particle size can be maintained in a narrow size range. Even though carbonyl powders are expensive, their ready commercial availability has made them the material of choice from the inception of the MIM process; they are largely responsible for MIM being a reasonably successful production process. With the addition of fine graphite, many steel compositions may be formulated and processed.

The second difficulty with fine metal powders is the very reason that they are used; their very high surface energy. This very high surface area per unit mass creates a major potential for instantaneous surface oxidation and a heat of reaction that could make the process self-sustaining and even explosive. The finer the metal powders, the greater the degree of pyrophoricity. Many metal powders in the range of 1 to 20 microns are reasonably stable, but this is related to the thermodynamic stability of the oxidation product. For example, fine metal powders of iron, nickel, copper, and molybdenum are stable, but those of aluminum, titanium, and zirconium are not, because of their strong affinity for oxygen. A protective atmosphere may be beneficial during storage of the metal powder; it is essential during any elevated temperature processing. In the specific case of oxide ceramics, pyrophoricity is not an issue since the powders are already in the form of the combustion product. Therefore, sub-micron size powders of oxide ceramics can be handled safely on a routine basis. Furthermore, thermal processing is performed under oxidizing atmospheres. Even nitrides or carbides are safe to handle as fine powders since they are relatively stable compounds. Of course, fine powders that become airborne have the potential to be inhaled by employees so that proper procedures must be implemented to guard against this possibility. Fine nickel powder is considered to be a carcinogen so that precautions should be taken to avoid dusting and, possibly, inhalation.

An understanding of the packing characteristics of the powders under study is essential in optimizing the MIM process[23]. Mono-sized particles have the least efficiency of packing and are therefore only of theoretical interest (unless a low density product is desired). Theoretically, an ordered close-packed structure (such as a face centered cubic, f.c.c., or close packed hexagonal, c.p.h., crystal) can a have a packing efficiency of about 74%, while a random packing of mono-sized spheres will have about 60% efficiency. A bimodal distribution requires a ratio of seven between the large and small particle diameters, for optimum packing. This is the specific case where the small spheres fill the interstices between the large spheres. In most commercially available powders one must expect a spectrum of particle sizes. The particle size distribution should be well characterized, and optimized for packing efficiency. Furthermore, this should be consistent from one lot of powder to another.

Generally, a spherical-shaped powder is preferred from the perspective of random packing and minimum binder content. Non-spherical powders are also acceptable, unless the length to diameter ratio is exceptionally large. In such cases, segregation and texturing can occur, resulting in anisotropic behavior in the green composite.

Two significantly different binder systems are in use at this time. The first is that developed by Wiech[10,11,12,24] and practiced by the Parmatech Corporation, in which a composite of metal and thermoplastic resin is heated and injected into a cold or warm die. The second is that described by Rivers[25]. Here, a cold feedstock consisting of the metal and a water

44

soluble binder is injected into a warm die where the rejection of moisture causes the mixture to solidify. The former process may require solvent extraction and a long (up to 120 hours) thermal binder removal and sintering step, while the latter requires only a few hours[26]. While offering the prospect of very short binder removal times, the methyl cellulose binder that is used in the Rivers' process tends to degrade readily and is considered to be explosive. This binder system and the accompanying technology is rarely practiced commercially.

While the details of the various binder systems are beyond the scope of this text, it is important to offer some guidance in order to understand the science of the process. In addition, it is important to recognize potential pitfalls that might lead to degradation of magnetic properties. The reader is encouraged to pursue further study of binder systems for MIM in the referenced literature.

A thermoplastic resin is most-often used as a carrier for the metal powders (e.g., polyethylene or polypropylene)[1,24]. Additional components are added to provide various attributes and yield a truly multi-component binder system[14,15]:

a) Other organics that degrade at higher or lower temperatures to promote binder degradation that is slow and sequential.
b) Organics that dissolve in selected solvents, for a two-stage binder removal system.
c) Dispersants that promote wetting of the powders by the binder, and a lowering of the composite viscosity.
d) Anti-oxidants that protect the metal powder from degrading.

A knowledge of these components and, especially their degradation products, is encouraged so that trace elements do not affect the final product. From an ecological and a safety perspective, the decomposition products of the binder system must be non-toxic.

One other important aspect of binder development is the technique that is used to combine or mix the binder components. It is essential that these components be mixed intimately and uniformly throughout. Furthermore, consistency from batch to batch is essential in dictating the reproducibility of the products fabricated. A variety of techniques such as mechanical mixing and "solution blending" (combination of all the ingredients in a solvent, followed by vaporization), can be used. During mechanical mixing, the optimum level of binder is indicated by a dramatic rise in the viscosity of the mixture[1,23].

After molding, the green compact may be subjected to an intermediate "solvent extraction" step to open up some pores or placed directly into an oven to initiate binder removal. Thermal degradation of the binder components usually proceeds slowly and must be performed in a controlled manner. This is perhaps the most difficult step in the MIM process since the removal of the support structure (the binder) results in a fairly weak porous framework that can be damaged easily. Much of any organic binder system will be removed by 538°C (1000°F). Very little dimensional shrinkage is observed up to this stage in ferrous-based alloy systems.

As the temperature is increased, atomic mobility increases and all the parameters discussed earlier come into effect. In the sintering stage, parts will exhibit shrinkages in the range of 15% to 25%, as the porous structure consolidates because of the high surface energies of the powders. The degree of shrinkage is very exact, if the process parameters are controlled well, and is dependent on the alloy system involved. Resultant densities greater than 98% are possible after a few hours at temperatures near 1316°C (2400°F). Note that conventional P/M shrinkages are usually less than 1%, and sintered densities are less than 92%. The presence of transient liquid phases will alter these significantly.

Beyond the sintering operation, the normal finishing operations (machining, grinding, plating, heat treating, etc.) may be used. Compared to conventional P/M, the MIM components will have reduced interconnected and surface porosity. This means that parts do not need to be resin-impregnated prior to plating or coating. Also, case hardening can be produced in MIM parts without intermediate operations such as copper infiltration. Note that case hardening and copper infiltration are not routinely performed on soft magnetic components since they degrade these properties.

The various practitioners of the MIM process have developed their own binder systems and regard this aspect of their technology as highly proprietary. Perhaps this is the only real identifiable difference between the various MIM part producers. The more astute practitioners will also focus efforts in advancing the technology from other perspectives, such as tooling design and development, alloy development, sintering cycle improvements, material handling systems (including automated handling), quality assurance systems, etc.

The debate on the effect of powder cost on the ability of the MIM process to compete effectively with other technologies still continues[27]. Although this is an important parameter, the significance of value-in-use pricing[28] should not be overlooked, especially for a technology that is so early in the business cycle. MIM does not compete with conventional P/M, but rather complements it. The MIM part can have significantly superior properties. If the only difference between a part being potentially P/M or MIM is few minor secondaries, perhaps the part should be processed by

P/M.

The cost factors are unlikely to be the same and a different approach must be used to advance this forming technique. The ability of MIM to produce a wide variety of shapes is well-recognized. However, an even greater value is the ability to process materials that are difficult, if not impossible, to produce by most other techniques. As an example for soft magnetic parts, almost any level of Si in Fe can be processed. While P/M is limited to less than 6% Si at best, because of the abrasive wear of tools and incompressibility of the powder mixture, MIM does not have these limitations. In MIM, any level of Si can be processed, including the 9% and 17% ferrosilicon masterblends. Any highly alloyed material that can not be compacted by conventional P/M is a good potential for MIM. In any case, for the same composition, a significant improvement in properties is realized by MIM over P/M, as will be discussed in Chapter 3.

Beyond this, opportunities exist for producing parts that would otherwise have been an assembly of several components. The point to be made is that the appropriate market must be identified and developed for MIM as against P/M. Perhaps a properly identified market should be allowed to dictate the value of the end product, rather than simply competing on a process cost basis alone.

REFERENCES

1. R. M. German, *Powder Injection Molding*, Metal Powder Industries Federation, Princeton, NJ, 1990.
2. W. D. Jones, *Fundamental Principles of Powder Metallurgy*, Edward Arnold, London, U.K., 1960.
3. F. V. Lenel, *Powder Metallurgy: Principles and Applications*, Metal Powder Industries Powder Federation, Princeton, NJ, 1980.
4. R. M. German, *Powder Metallurgy Science*, Metal Powder Industries Federation, Princeton, NJ, 1984.
5. E. Klar, Editor, *Powder Metallurgy*, Metals Handbook, Ninth Edition, Vol. 7, ASM International, Materials Park, OH, 1984, pp. 23-175.
6. J. K. Beddow, "The Production of Metal Powders by Atomization", *Monographs in Powder Science and Technology*, Heyden and Son Ltd., London, UK, 1978.
7. A. Lawley, *Atomization: The Production of Metal Powders*, Metal Powder Industries Federation, Princeton, NJ, 1992.
8. M. F. Ashby, "A First Report on Sintering Diagrams", *Acta Met.*, 1974, Vol. 22, pp. 275-289.
9. C. Lall, "Fundamentals of High Temperature Sintering: Application to Stainless Steels and Soft Magnetic Alloys", *Int. J. Powder Metall.*, 1991, Vol. 27, No. 4, pp. 315-329.

10. R. E. Wiech, Jr., "Manufacture of Parts from Particulate Materials", U.S. Patent 4,197,118, (8 April, 1980).
11. R. E. Wiech, Jr., "Method for Removing Binder from a Green Body", U.S. Patent 4,404,166, (13 September, 1981).
12. R. E. Wiech, Jr., "Method of Making Inelastically Compressible Ductile Particulate Material Article and Subsequent Working Thereof", U.S. Patent 4,445,936, (1 May, 1984).
13. K. Schwartzalder, "Injection Molding of Ceramic Materials", *Bulletin of American Ceramic Society,* 1949, Vol. 28, No. 11, pp. 459-461.
14. D. W. Richerson, *Modern Ceramic Engineering: Properties, Processing and use in Design,* Marcel Dekker, New York, NY, 1982.
15. J. A. Mangels and G. L. Messing, "Forming of Ceramics", *Advances in Ceramics,* Vol. 9, American Ceramic Society, Columbus, OH, 1984.
16. C. Lall, "Injection Molding of Advanced Structural Ceramic Components", DuPont Experimental Station laboratory, Internal Report, 1986.
17. J. A. Mangels and W. Trela, "Ceramic Components by Injection Molding", *Advances in Ceramics,* Vol. 9, "Forming of Ceramics", J. A. Mangels and G. L. Messing (Eds.), American Ceramic Society, Columbus, OH, 1984, pp. 220-233.
18. K. Hunold, J. Greim and A. Lipp, "Injection Molded Ceramic Rotors - Comparison of SiC and Si_3N_4", *Powder Metall. Int.,* 1989, Vol. 21, No. 4, pp. 17-23.
19. L. F. Pease, III, "Present Status of P/M Injection Molding, (MIM), - An Overview", *Progress in Powder Metall.,* 1987, Vol. 43, pp. 789-828.
20. L. F. Pease, III, "Metal Injection Molding - Overview, Process/Industry", MIMA Symposium, (Indianapolis, IN), organized by MPIF, Princeton, NJ, (November 16, 1989).
21. L. W. Baum, 1992, Remington Arms Company (DuPont), Ilion, NY, private communication.
22. A. J. Neupaver and R. P. Mason, "Metal Powder for Injection Molding", K. H. Moyer (Ed.), Metal Injection Molding Seminar Proceedings, Metal Powder Industries Federation, Princeton, NJ, 1988.
23. R. M. German, *Particle Packing Characteristics,* Metal Powder Industries Federation, Princeton, NJ, 1989.
24. A. R. Erickson and R. E. Wiech, Jr., "Injection Molding", E. Klar (Ed.), "Powder Metallurgy", Metals handbook, Ninth Edition, Vol. 7, ASM International, Materials Park, OH, 1984, pp. 495-500.
25. R. Rivers, U.S. Patent 4,113,480, (1978).
26. E. Andreotti, "Metal Injection Molding Experience at P/M Plant", *Industrial Heating,* pp. 18-20, (January, 1988).
27. P. U. Gummeson, "The Metal Injection Molding Opportunity- A Critical Review", *Int. J. Powder Metall.,* 1989, Vol. 25, No.3, pp. 207-216.
28. R. C. Drewes, "Cost Effective Production of MIM Components", *Advances in Powder Metallurgy-1991,* Vol. 2, compiled by L. F. Pease and R. J. Sansoucy, Metal Powder Industries Federation, Princeton, NJ, pp. 297-306.

III.

SOFT MAGNETIC PROPERTIES OF SELECTED ALLOYS

The soft magnetic performance of a component is dependent on its composition and processing history. In this chapter, the magnetic properties of materials processed by powder metallurgy and metal injection molding are discussed in detail. The influence of processing parameters and the chemistry of the part on the soft magnetic properties is highlighted.

3.1 Introduction

The specific properties of materials made by powder metallurgy and metal injection molding are dictated by the degree to which the practitioners of these manufacturing methods follow the fundamentals of their science. These properties are described for selected materials and compared to each other, as well as their wrought counterparts. Furthermore, the impact of the processing conditions on these properties is illustrated with specific examples.

The majority of the information discussed in this chapter was developed using test procedures similar to those described by Lall and Baum[1]. Except where otherwise stated, all powder metal blends used a high purity, commercially available, iron base powder. Fe_3P was used to make the Fe-P alloys; while a ferro-silicon masteralloy, containing 17% Si, was used for the Fe-Si alloys. The Fe-Ni alloys were modified from the base prealloyed Fe-50% Ni powder with appropriate elemental additions of Fe or Ni. The ferritic stainless steel powders were prealloyed and in the annealed condition. Both zinc stearate and acrawax C lubricants were employed; compaction pressures were in the range 414 to 828 MPa (30-60 tons per square inch). Sintering was generally for about 45 - 60 minutes at temperatures ranging from 1121°C (2050°F) to 1371°C (2500°F). A variety of protective atmospheres such as dissociated ammonia (DA), hydrogen, nitrogen, and vacuum were used.

Much of the magnetic data were obtained by the hysteresis method mentioned briefly in Section 1.10.2 and described more fully in ASTM standard test methods (see Appendix B). The edges of the toroidal-shaped samples were deburred and dimensional measurements made. Primary and secondary coils were wound over a layer of insulation tape.

The test equipment used for measuring magnetic properties is shown in Figure 3.1, and the electrical circuit diagram is shown in Figure 3.2. The value of the applied field, H_{max}, was calculated using Equation [1.12], where L is replaced by the average of the inner and outer circumferences. A current of 1 ampere was used to calibrate the X-axis of the X-Y recorder. Tests were normally performed using an applied field of at least 1990 A/m (25 Oe). A Function Generator was used to drive a Bipolar DC Power Supply and the results plotted automatically on the X-Y Recorder. For higher frequencies, a trace was retained in a Digital Storage Oscilloscope and then transferred electronically to the X-Y recorder. Different expansions of the X (i.e., H axis) were made so that H_c could be read more accurately. In addition, the cycle rate was slowed to 0.0001 hertz near H=0, for the same reason.

It is important to be clear on the method of testing that is used to

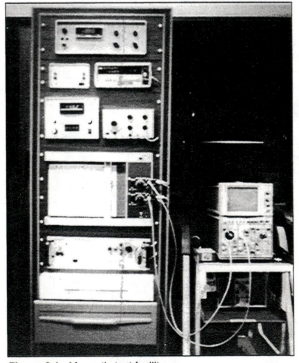

Figure 3.1: Magnetic test facility.

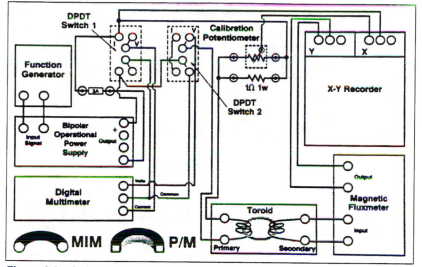

Figure 3.2: A circuit diagram for test facility. Insert shows different types of cross-sections for the MIM and P/M rings.

measure magnetic properties, since erroneous conclusions can be made by comparing dissimilar test data. The test method outlined above was used to test three standard ASTM toroids that had been evaluated in a "round robin" test program. The information that is maintained with these standard samples is reproduced in Table 3.1. Values obtained by this study using the test method described earlier are shown in Table 3.2. An inspection of the data shows that the present test measurements obtained from the described equipment are consistent with those documented by this internationally recognized standards organization. Table 3.3 shows the effect of using different values of the applied field, H_{max}. The effect of changing this parameter is to alter the values of B_r and B_m. Therefore, it is important to state the maximum value of the magnetizing field, H, when reporting magnetic data. Both coercive field and maximum permeability appear to be relatively unaffected.

Table 3.1 Data Supplied with ASTM Standard Toroids (1979)

	TOROID NUMBER		
	67004	67007	67010
MATERIAL:	GRAIN OR. M-5 Fe-3% Si	NON-OR. M-36 Fe-1.4% Si	NON-OR. M-15 Fe-4% Si
DENSITY (g/cm^3)	7.65	7.75	7.65
MAG. PATH L (cm)	35.91	35.91	35.91

Table 3.1 (Cont.) Data Supplied with ASTM Standard Toroids (1979)

	TOROID NUMBER		
	67004	67007	67010
MATERIAL:	GRAIN OR. M-5 Fe-3% Si	NON-OR. M-36 Fe-1.4% Si	NON-OR. M-15 Fe-4% Si
ACTIVE WT (g:lbs)	414:0.9127	428:0.9436	414:0.9127
SAMPLE AREA (cm^2)	1.507	1.538	1.507
THICKNESS (in)	0.012	O.0185	0.014
PRIMARY TURNS	265	265	265
SECONDARY TURNS	265	265	265
PRIMARY WINDING RESISTANCE (ohm)	0.371	0.371	0.370
FLUX VOLTS AT 10 GAUSS, 60 Hz	10.64	10.86	10.64

Table 3.2 Measured Data for ASTM Toroids (After Lall and Baum[1])

	TOROID NUMBER		
	67004	67007	67010
MATERIAL:	GRAIN OR. M-5 Fe-3% Si	NON-OR. M-36 Fe-1.4% Si	NON-OR. M-15 Fe-4% Si
ASTM DATA AT H = 796 A/m (1.078 ampere)			
Average of all B_m values, tesla	1.8423	1.5773	1.4182
AT CALCULATED H = 738 A/m (1.00 ampere)			
B_m (tesla)	1.813	1.550	1.425
B_r (tesla)	1.588	1.400	1.075
H_c (A/m)	11.14	43.77	25.47
μ_r	48,000	12,000	13,000
AT CALCULATED H = 1194 A/m (1.6 ampere)			
B_m (tesla)	1.850	1.588	1.463
B_r (tesla)	1.588	1.406	1.075
H_c (A/m)	11.94	43.77	25.46
μ_r	45,000	11,000	13,000
AT CALCULATED H = 1512 A/m (2 ampere)			
B_m (tesla)	1.886	1.600	1.475
B_r (tesla)	1.588	1.406	1.075
H_c (A/m)	11.94	44.57	25.47
μ_r	40,000	11,000	13,000

Table 3.3 Showing Effect of Different Applied Fields (after Lall[20]).

SOURCE	PROPERTY				
	DENSITY	B_m	B_r	μ_r	H_c
	(kg/m^3)	(T)	(T)		(A/m)
Pure iron:					
At H = 1194 A/m	7000	1.09	0.97	2400	135
At H = 1990 A/m	7000	1.15	1.00		135
At H = 7958 A/m	7000	1.35	1.10		135
Fe-0.45% P:					
At H = 1194 A/m	7000	1.19	1.12	3300	127
At H = 1990 A/m	7000	1.24	1.14		127
At H = 7958 A/m	7000	1.43	1.24		127
Fe-50% Ni:					
At H = 1194 A/m	7000	0.99	0.81	12,300	22.3
At H = 1990 A/m	7000	1.01	0.84		22.3
At H = 7958 A/m	7000	1.14	0.88		22.3

NOTE: All samples sintered at 1249°C, 45 min., in DA.

In the following sections, the format used by Lall and Baum[1] for representing data will be adopted. The figures will exhibit the effect of sintering temperature for a given alloy system. In addition, the data will be transposed to tables and supplemental data from the published literature added to the table. Where significant differences in data or process procedures are known to exist, an effort will be made to provide explanations for the variations. The intent is to demonstrate the effects of various practical parameters on the soft magnetic performance of P/M and MIM materials. Note that the values in the figures can be regarded as being typical properties for a given material, while the values in the tables will relate the ranges possible.

3.2 Iron

The intrinsic saturation induction (B_s-H) of ferromagnetic iron is about 2.16 tesla (21.6 kilogauss), which is the highest value of any of the elements in the pure form. Cobalt, at 1.79 tesla (17.9 kG), has the second highest value. Only the combination of these two elements together with a third element (2% V, which improves ductility) produces vanadium

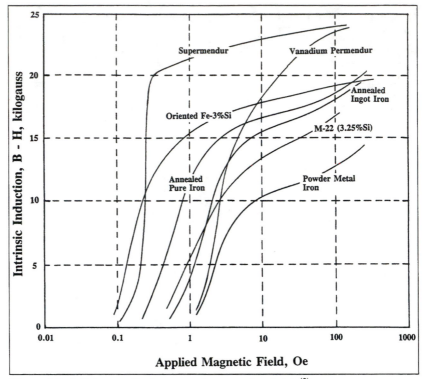

Figure 3.3: Initial magnetization curves for several materials[2].

permendur, which has a higher saturation induction value (about 2.40 tesla [24.0 kG])[2]. Figure 3.3 shows the initial magnetization curves for a variety of materials. An important point to recall is that in many soft magnetic applications, the applied field rarely exceeds 1600 A/m (about 20 Oersteds).

The introduction of any defects (e.g., vacancies, lattice strains, grain boundaries, dislocations, pores, impurities, etc.) into iron invariably result in a <u>decrease</u> in the induction values. It is important to recall that the highest value of this property is only realized with these elements in their purest state and under very stringent laboratory conditions. For polycrystalline iron in a reasonably practical environment, the saturation induction for iron is significantly lower and will decrease as the level of contaminants increase. Table 3.4 shows the magnetic properties for wrought iron from Figure 3.3 and as specified by ASTM material standard A-848. The maximum induction is about 1.7 tesla (17 kG) near H = 1592 A/m (20 Oe), while the coercive field is 80 A/m (roughly 1.0 Oe).

Table 3.4 Magnetic Data on Pure Iron

SOURCE	PROPERTY				
	DENSITY	B_m	B_r	μ_r	H_c
	(kg/m^3)	(T)	(T)		(A/m)
WROUGHT					
ASTM A 848	7860	1.70	1.28	5,000	79.58
(Electrical resistivity 0.13 micro-ohm·m)					
POWDER METAL					
1. ASTM A811-87	6600	0.90	0.76	1800	175
H=1194 A/m	6900	1.06	0.91	2100	159
	7200	1.23	1.07	2500	159
2. Lall & Baum[1]: H = 1990 A/m					
1260°C, DA	6800	1.14	0.96	2900	131
45 min.	7200	1.36	1.10	3700	127
	7400	1.47	1.29	4700	119
(Resistivity 0.14, 0.12, 0.11 micro-ohm·m at 6800, 7200, and 7400 kg/m^3, respectively).					
3. McDermott[21]: H = 1194 A/m					
1121°C, DA	6700	0.98	0.83	1988	161
30 min.	7000	1.14	0.97	2317	165
	7200	1.24	1.06	2651	166
4. McDermott[21]: H = 1194 A/m					
1121°C, H$_2$	6800	1.06	0.92	2534	163
30 min.	7100	1.21	1.07	3069	163
	7300	1.29	1.16	3337	155
5. McDermott[21]: H = 1194 A/m					
1260°C, DA	6700	1.00	0.77	1818	146
30 min.	7000	1.17	0.91	2178	146
	7300	1.27	1.05	2794	146
6. McDermott[21]: H = 1194 A/m					
1260°C, H$_2$	6900	1.10	0.95	2727	140
30 min.	7100	1.25	1.09	3189	137
	7300	1.33	1.17	3566	134

Table 3.4 (Cont.) Magnetic Data on Pure Iron

SOURCE	PROPERTY				
	DENSITY	B_m	B_r	μ_r	H_c
	(kg/m^3)	(T)	(T)		(A/m)
7. Gagne et al.[22]: H = 1194 A/m					
1120°C, H$_2$	7400	1.24	1.16	2900	135
1260°C, H$_2$	7400	1.25	1.20	3600	119
8. Lall[20]: H = 1990 A/m					
1260°C, vacuum	7000	1.17	1.09	3500	111
9. Moyer[23]: H = 1194 A/m					
1121°C, H$_2$ + N$_2$	7000	1.14	1.10	2800	143
Shot-peened	7000	0.97	0.87	1700	159
10. Mossner[25]					
1121°C,	7000	1.00	0.89	2000	159
1288°C	7000	1.07	0.94	2400	127

Table 3.4 also shows the soft magnetic properties of pure iron processed by powder metallurgy, as related by a number of sources. The most notable difference between wrought and P/M iron is the lower induction values for the latter. This is directly related to the increased level of porosity in the P/M product. The empirical relationship as given by Moyer, et al.[3] and Adler, et al.[4] is:

$$B_S/B_{SN} = 1 - aP \qquad [3.1]$$

Here, B_S and B_{SN} are the saturation induction values for porous and non-porous materials, respectively. "P" is the porosity level and "a" is a constant having values ranging from 1.5 to 2. Adler, et al.[4] refined this further and reported that the flux density is linearly related to the fractional density of the material:

$$\frac{B_S}{B_{SM}} = (1 - aP)^n = (\frac{d}{d_M})^n \qquad [3.2]$$

where   d is the apparent density
          d_M is the pore-free density
          B_{SM} is the pore-free flux density
          n has a value of about 1.5

Figure 3.4 shows the effect of sintering temperature on the soft magnetic properties of pure iron[1]. The higher sintering temperature causes an increase in sintered density and an improvement in the microstructural cleanliness. This results in the observed improvements in soft magnetic properties.

Figure 3.5 shows an optical micrograph of pure iron sintered at 1260°C in a vacuum furnace. For this sample, and others to follow, the vacuum sintering was for about 45 minutes in the high heat section, with a backfill of pure nitrogen to a pressure of about 750 microns of mercury. Also, nitrogen was used to quench the parts in a separate chamber. This system

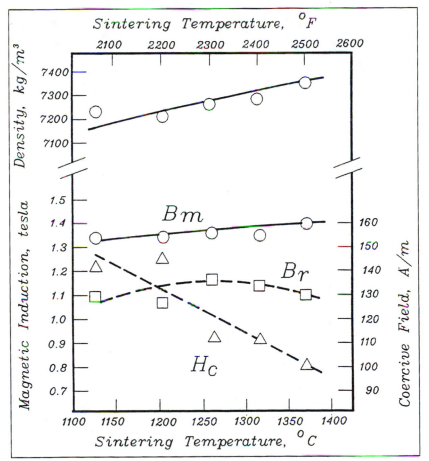

Figure 3.4: Magnetic properties of pure Fe as a function of sintering temperature in dissociated ammonia[1].

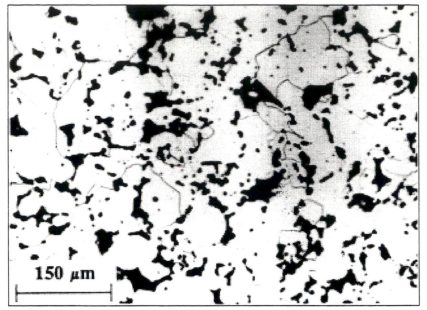

Figure 3.5: Microstructure of pure iron sintered in vacuum at 1260°C for 45 minutes. 100% ferrite matrix with typical porosity at sintered density of 7000 kg/m³.

was effective in cooling the parts below about 550°C in less than 2 minutes. The microstructure in Figure 3.5 for iron, shows the ferrite matrix and a typical pore structure for a pressed and sintered product.

Iron parts made by powder metallurgy are often limited to use in direct current (d.c.) type applications because of their high coercive field and low electrical resistivity. These result in slow response times and high energy losses during alternating cycle (a.c.) type applications. Iron parts made by P/M are normally used in applications in place of wrought iron or low carbon steels (e.g., AISI 1008 or 1010), and are frequently acting as magnetic flux carriers. In this function, the most important magnetic parameter is maximum induction, B_m.

For iron parts, density has the greatest controlling influence on magnetic properties. As shown in Figure 3.4, the density increases as the sintering temperature is increased. Another way to improve density is to use high compressibility powders, which are essentially high-purity, well-annealed materials. However, even these must be sintered reasonably well for optimum soft magnetic properties. Very often, iron and steel P/M parts are sintered at 1121°C (2050°F) for 30 minutes or less, which means that there is very little densification. This low temperature is beneficial from the point of maintaining dimensional control, since there is very little shrinkage.

However, from a metallurgical and high performance magnetics standpoint, this leads to deficiencies in that elements have not diffused properly, grain size is still small, inclusions (e.g., oxides, carbides, nitrides) have not been dissolved, etc. Use of fine powders aids the lower temperatures in attaining higher densities. The range of values shown in Table 3.4 clearly demonstrates the impact of such processing variables, since each source of the data developed their own preferred method of processing.

Improved soft magnetic properties are obtained by sintering at a high temperature in a vacuum or a 100% pure hydrogen atmosphere. A concern is that nitrogen absorption from a DA atmosphere may degrade soft magnetic properties. This will be discussed further in sub-section 3.9.3 on impurity effects.

3.3 Iron-Phosphorus

The Fe-P system has been promoted commercially in recent years as a new soft magnetic material that is claimed to replace virtually all other high performance soft magnetic P/M materials[5]. However, the fact is that Fe-P P/M alloys have been around for a long time[6] and their primary contribution to the P/M industry is that they enable the parts producer to sinter them near 1121°C (2050°F). Excellent studies[7,8,9] on the effect of P on Fe show that the element stabilizes the more open b.c.c. phase and forms a liquid phase above 1050°C (1922°F). Both of these facts lead to a high rate of sintering at the commonly used temperature of 1121°C (2050°F).

Fe-0.45% P is the most common composition in this alloy system in use in the P/M industry, both in Europe and the U.S.A. While reasonable magnetic properties are achieved by sintering at 1121°C (2050°F), considerable improvements are seen by sintering at higher temperatures (Figure 3.6) and longer times[7,9]. Apart from magnetic property improvements, very marked improvements in ductility are also realized (Figure 3.7).

Figure 3.6 shows the dramatic effect of sintering temperature on the soft magnetic properties of Fe-0.45% P. Table 3.5 displays these parameters as measured by various workers for this alloy system. Lindskog, et al.[7] also showed that the addition of 0.45% P to plain Fe causes an increase in B_m (about 1.25 to 1.35 tesla [12.5 to 13.5 kG]), a decrease in coercive field (151 to 103 A/m [1.9 to 1.3 Oe]) and a rise in resistivity (0.13 to 0.20 micro-ohm·m). These are positive changes and due primarily to the phase stabilization and liquid phase sintering. The resistivity increase, due to the alloying effect of P, and the lower H_c lead to a reported[7] drop in total energy losses from about 24 watts/kg for Fe to about 16 watts/kg for Fe-0.45% P. Note that Fe-P alloys cannot be fabricated by the conventional

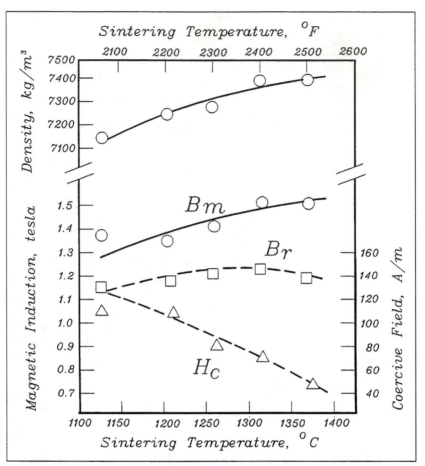

Figure 3.6: Magnetic properties of Fe-0.45%P as a function of sintering temperature in dissociated ammonia[1].

wrought metallurgy route, since the P results in "hot-shortness" during the elevated temperature rolling operation.

The major drawback to high temperature sintering for the Fe-P alloy system is the unusually high shrinkage[7,8] (Figure 3.7). The same difficulty is seen[9] when processing Fe-0.80% P, which has improved magnetic properties over Fe-0.45% P. In such cases, a coining operation is often necessary to control dimensions. A subsequent low temperature magnetic anneal is highly recommended to remove the surface deformation without altering dimensional control. In high performance a.c. soft magnetic applications, such an anneal is almost mandatory as the surface region is

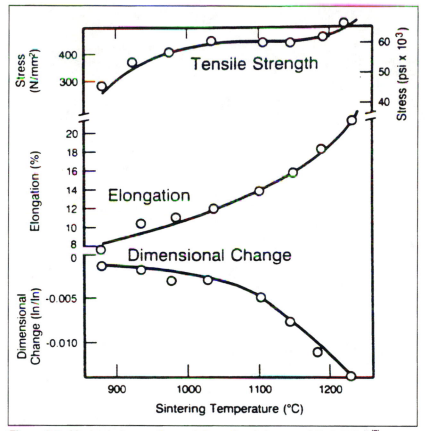

Figure 3.7: Mechanical properties and dimensional changes of Fe-0.45% P[7].

the only area that is magnetized and demagnetized in each cycle.

Figure 3.8 shows a well-sintered microstructure for Fe-0.45 %P processed in vacuum at 1260°C. The well-rounded porosity and relatively high density (7300 kg/m^3), are signs of a well-sintered structure. The matrix is all ferrite. The transient liquid phase in the Fe-P system is the primary reason for the differences in microstructure between Figures 3.5 and 3.8. Both alloys were processed in an identical manner.

Table 3.5 illustrates the value of high temperatures and vacuum or hydrogen atmospheres in offering improved properties. The impact of C-O-N absorption will be covered in sub-section 3.9.3, below.

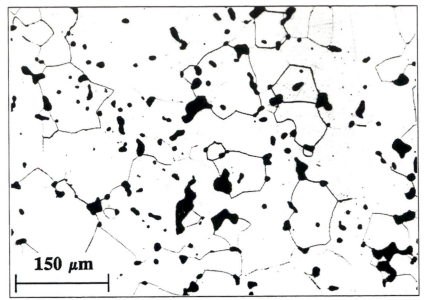

Figure 3.8: Microstructure of Fe-0.45% P sintered in vacuum at 1260°C for 45 minutes. Shows 100% ferrite matrix with well-rounded porosity. Sintered density of 7300 kg/m³.

Table 3.5 Magnetic Data on Fe-0.45% P

SOURCE	PROPERTY				
	DENSITY	B_m	B_r	μ_r	H_c
	(kg/m³)	(T)	(T)		(A/m)

WROUGHT
NOT AVAILABLE IN WROUGHT FORM.

POWDER METAL

1. ASTM A-839	6800	1.07	0.87	2400	135
	7100	1.19	0.99	2800	135
	7200	1.27	1.08	3100	127

2. Lall and Baum[1]: H = 1990 A/m

1260°C, DA	7000	1.23	0.99		96
45 min.	7200	1.34	1.12	4800	80
	7400	1.46	1.26		60

(Resistivity 0.23, 0.21, 0.20 micro-ohm·m at 7000, 7200, and 7400 kg/m³, respectively).

Table 3.5 (Cont.) Magnetic Data on Fe-0.45% P

SOURCE	PROPERTY				
	DENSITY	B_m	B_r	μ_r	H_c
	(kg/m^3)	(T)	(T)		(A/m)
3. McDermott[21]: H = 1194 A/m					
1120°C, DA	6840	1.11	0.97	2831	139
30 min.	7120	1.24	1.10	3286	139
	7270	1.32	1.18	3701	163
4. McDermott[21]: H = 1194 A/m					
1260°C, DA	7150	1.27	1.10	4216	110
30 min.	7350	1.39	1.35	4073	106
	7450	1.39	1.18	4445	103
5. Gagne et al.[22]: H = 1194 A/m					
1120°C, H_2, 30 min.	7400	1.26	1.23	4300	111
1260°C, H_2, 30 min.	7400	1.32	1.28	5000	96
6. Moyer[23]: H = 1194 A/m					
1121°C	7200	1.22	1.10	3200	140
1288°C	7200	1.32	1.18	4800	104
7. Lall[20]: H = 1990 A/m					
1260°C, vac.	7200	1.35	1.25	6000	80
8. Moyer[23]: H = 1194 A/m					
1121°C, $H_2 + N_2$	7100	1.22	1.19	3600	119
Shot-peened	7100	0.99	0.87	2000	151
9. Mossner[25]					
1121°C,	7200	1.02	0.87	2250	143
1288°C	7400	1.31	1.15	5550	72

3.4 Iron-Silicon

As Figure 3.9 shows, magnetic properties degrade very quickly if sintering temperatures below 1200°C (about 2200°F) are used. The stable silicon oxides require high temperatures to dissociate and the transient liquid phase sintering near 1260-1316°C (2300-2400°F) leads to rapid densification and, therefore, improvements in both mechanical and magnetic properties. Table 3.6 also lists additional values from other sources.

In this alloy system, the reduction of the unusually stable silicon oxides is the most difficult task during sintering. High temperatures and low dew points, together with high hydrogen contents, aid in the reduction of the oxides[10,11]. The other major factor that assists sintering of Fe-Si alloys is

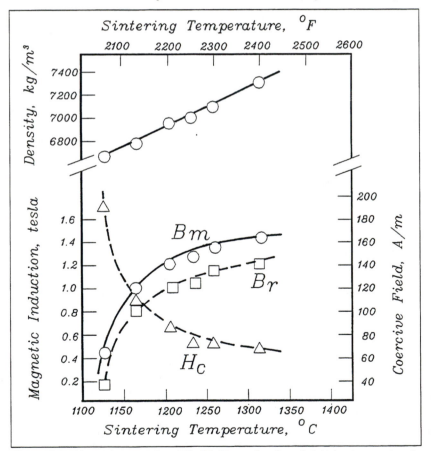

Figure 3.9: Magnetic properties of Fe-3% Si as a function of sintering temperature in dissociated ammonia[1].

the transient liquid phase. As for the Fe-P alloy system, this liquid phase is responsible for the high density product shown in Figure 3.10. This optical micrograph displays a ferritic single-phase structure and well-rounded pores.

Table 3.6 Magnetic Data on Fe-3% Si

SOURCE	PROPERTY				
	DENSITY	B_m	B_r	μ_r	H_c
	(kg/m^3)	(T)	(T)		(A/m)

WROUGHT
ASM Handbook[2] 7650 20.1(B_s) - 8,000 56
(3% Si steel has electrical resistivity of 0.47 micro-ohm·m).

Carpenter Technology[27] 7650 1.6 0.6 5,000 56
(2.5% Si steel, "B-FM" grade, B_m at H = 796 A/m, B_s = 2.06 T, resistivity 0.40 micro-ohm·m)

Carpenter Technology[27] 7600 1.5 0.4 4,000 48
(4% Si steel, "C" grade, B_m at H = 796 A/m, B_s = 2.0 T, resistivity 0.58 micro-ohm·m)

POWDER METAL
1. Lall and Baum[1]: H = 1990 A/m

1260°C, DA	6800	1.17	0.94	2900	104
45 min.	7000	1.31	1.09	3700	92
	7200	1.39	1.18	4900	80

(Resistivity 0.59, 0.55, 0.52 micro-ohm·m at 6800, 7000, and 7200 kg/m^3, respectively).

2. Lall[20]: H = 1990 A/m

1260°C, vac.	7200	1.28	1.17	7000	64

3. Mossner[25]: H = 1194 A/m

1121°C,	6700	0.99	0.68	1750	143
1260°C	7400	1.19	0.91	4050	72

4. Moyer and Ryan[26]: H = 1194 A/m

1232°C, vac. 1 hr.	7200	1.35	1.21	7500	72
1232°C, vac. 1 hr.	7400	1.23	0.75	7000	48

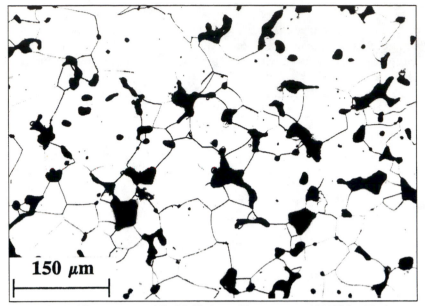

Figure 3.10: Microstructure of Fe-3% Si sintered in vacuum at 1260˚C for 45 minutes. Shows ferrite matrix with well-rounded porosity. Sintered density of 7100 kg/m³.

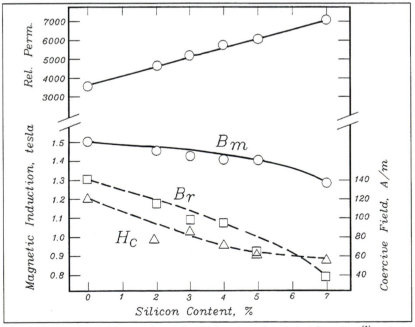

Figure 3.11: Magnetic properties of Fe-Si alloys as a function of Si content[1].

Figure 3.11 shows the effect of Si additions on the magnetic properties of Fe. An additional benefit is that the resistivity increases from about 0.13 for Fe, to about 0.55 micro-ohm.m for Fe-3% Si. The latter composition is most popular in commercial practice, since it is a compromise between the lowering B_m and increasing resistivity as Si content is increased. Furthermore, Si additions result in a significant decrease in ductility. A total energy loss of about 5 watts/kg is expected in Fe-3% Si, which is about one-third of the value for Fe-0.45% P. Higher Si alloys up to about 6% Si can be fabricated, with difficulty, into simple shapes. These higher Si alloys result in greater tool wear and a lower induction. Such compositions are very difficult to fabricate from wrought bar stock because of their inherently low ductility. In commercial practice, 3.5% is about the highest level of silicon that is used for wrought products, because of the ductility issue.

3.5 Iron-Nickel

Figure 3.12 shows the effect of processing temperature, while Figure 3.13 shows the effect of nickel content on the magnetic properties of Fe. While the sintering temperature has some effect on the soft magnetic properties, it is not as critical as in the case of the Fe-P or Fe-Si systems. The most significant effect of increasing the nickel content is a drop in B_r and a rise in H_c. Table 3.7 shows additional data at different densities.

Figure 3.14 shows an optical micrograph of Fe-50% Ni. The matrix consists of single phase austenite with annealing twins. The pores are well-rounded and the presence of annealing twins are signs of a well-sintered product. Since the Fe-Ni alloy system does not have the benefit of a transient liquid phase to aid material transfer, the sintered density is lower than that for the Fe-P and Fe-Si systems. A high temperature sinter to aid diffusion of the substitutional elements will improve density. Another approach is to simply repress the sintered product. This should be followed by an anneal to take care of the deformation damage, otherwise a poor soft magnetic response will be exhibited.

Adler, et al.[4] demonstrated the dependence of the soft magnetic properties of this alloy system on the porosity, inclusion level and grain size. The flux density essentially follows the relationship given by Equation [3.2]. The demagnetizing fields as a result of the porosity are thought to be responsible for the greater than linearity dependence. If the size of the inclusions is close to the domain wall thickness (about 0.03 to 0.2 μm), the effect on structure sensitive properties such as H_c is essentially linear with respect to inclusion content. The inclusions will interact most effectively with the Bloch walls when their sizes are similar. In addition, H_c bears a linear relationship to the reciprocal of the grain size.

It is clear that coercive field (H_c) and residual induction (B_r) are much improved in the Fe-50% Ni alloy over the Fe-3% Si alloy. Admittedly the B_m values are lower for Fe-50% Ni, but overall this would be a far superior material for high performance soft magnetic a.c. applications–especially when one realizes that the maximum permeability for Fe-50% Ni is about four times that for Fe-3% Si and almost six times that for pure Fe.

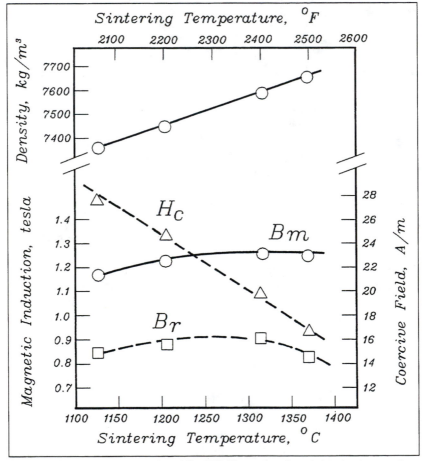

Figure 3.12: Magnetic properties of Fe-50% Ni as a function of sintering temperature in dissociated ammonia[1].

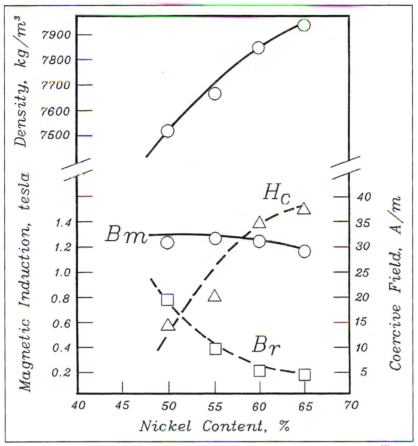

Figure 3.13: Magnetic properties of Fe-Ni alloys as a function of nickel content[1].

Table 3.7 Magnetic Data on Fe-50% Ni

SOURCE	PROPERTY				
	DENSITY	B_m	B_r	μ_r	H_c
	(kg/m^3)	(T)	(T)		(A/m)

WROUGHT
ASM Handbook[2]   8200   1.60   0.80   70,000   3.98
("hypernik" has an electrical resistivity of 0.50 micro-ohm·m).

69

Table 3.7 (Cont.) Magnetic Data on Fe-50% Ni

SOURCE	PROPERTY				
	DENSITY	B_m	B_r	μ_r	H_c
	(kg/m^3)	(T)	(T)		(A/m)

POWDER METAL

1. Lall and Baum[1]: H = 1990 A/m

1260°C, DA	6800	0.93	0.71		20.7
45 min.	7100	1.09	0.80		19.9
	7500	1.27	0.94	21000	19.1

(Resistivity 0.78, 0.69, 0.60 micro-ohm·m at 6800, 7100, and 7500 kg/m^3, respectively).

2. Lall[20]: H = 1990 A/m

1260°C, vac	7400	1.08	0.86	10600	23.9

3. Mossner[25]

1121°C	7100	1.15	0.85	11000	31.4
1288°C	7300	1.29	0.90	16000	23.9

4. Moyer and Ryan[26]: H = 1194 A/m

1246°C, vac. 2 hr.	7300	1.12	0.70		23.9
1246°C, vac. 2 hr.	7500	1.23	0.75		23.9

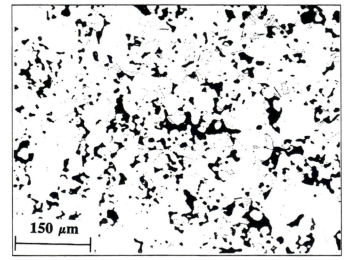

150 μm

Figure 3.14: Microstructure of Fe-50% Ni sintered in vacuum at 1260°C for 45 minutes. Shows single phase austenite with annealing twins. Sintered density of 7200 kg/m^3.

3.6 Ferritic Stainless Steels

The ferritic stainless steels became popular in the latter part of the 1980's for the "tone-wheels" or "sensor-rings" used in Anti-lock Brake Systems (ABS) because of their magnetic response and their moderate corrosion resistance. Another factor was the ability to develop a considerable degree of ductility by vacuum sintering these stainless steels[12,13]. Since many of the assembly operations required the wheel to be forced onto the wheel axle and thereby form a tight mechanical lock, ductility was a key property that had to be met.

If the ferritic stainless steels (e.g., AISI grades 410L, 420L, 434L, etc.) are sintered in conventional nitrogen-containing atmospheres, very little ductility is observed, so they can not be used in such applications. Furthermore, the magnetic properties (Table 3.8) and corrosion resistance deteriorate with the absorption of nitrogen[14]. These ferritic stainless steels become martensitic with the absorption of carbon or nitrogen[14]. In the martensitic form, the product is very hard and brittle.

Figure 3.15 shows well-sintered microstructures for 410L and 434L obtained by sintering at 1260°C in vacuum. Both stainless steels are ferritic as long as the level of C and N_2 is kept low[14], preferably below 0.01%.

Table 3.8 Magnetic Data on Ferritic Stainless Steels

SOURCE	PROPERTY				
	DENSITY	B_m	B_r	μ_r	H_c
	(kg/m^3)	(T)	(T)		(A/m)
WROUGHT					
Carpenter Technology[27]	7620	1.2	0.6	2000	159
(430F, B_m at H = 796 A/m, B_s = 1.42 T)					
POWDER METAL					
410L					
1. Lall[20]: H = 1990 A/m					
1260°C, Vac. 45 min.	7100	1.09	0.94	2200	169
1288°C, H_2	7100	5.6	-	320	590
430L					
2. McDermott[21]: H = 1194 A/m					
1121°C, H_2	6450	0.73	0.47	1000	182
30 min.	6670	0.79	0.51	1043	185

Table 3.8 (Cont.) Magnetic Data on Ferritic Stainless Steels

SOURCE	PROPERTY				
	DENSITY	B_m	B_r	μ_r	H_c
	(kg/m^3)	(T)	(T)		(A/m)
3. McDermott[21]: H = 1194 A/m					
1121°C DA	5810	0.034	0.003	11	231
30 min.	6130	0.041	0.005	13	294
	6420	0.045	0.005	14	279
4. Moyer and Jones[13]: H = 1194 A/m					
1260°C, H_2	7250	1.05	0.80	1900	159
5. Svilar and Ambs[14]: H = 1194 A/m (vac + backfill of listed gas)					
1121°C, H_2	6690	0.81	0.74	1200	207
1232°C, H_2	7110	0.98	0.89	1800	167
1121°C, $H_2 + N_2$	6490	0.016	0.009		740
1232°C, $H_2 + N_2$	6990	0.038	0.024		669
434L					
6. Lall[20]: H = 1990 A/m					
1260°C, Vac. 45 min.	7000	1.01	0.84	1700	159
1288°C, H_2	7100	6.5	-	450	220
7. McDermott[21]: H = 1194 A/m					
1288°C, DA, 30 min.	6430	0.728	0.463	1092	160
1288°C, H_2, 30 min.	6650	0.791	0.483	1165	151
8. McDermott[21]: H = 1194 A/m					
1121°C, DA	5830	0.053	0.009	416.8	414
30 min.	6030	0.063	0.013	719.7	477
	6290	0.079	0.019	424.9	533
9. Mossner[25]					
1288°C	7100	0.89	0.44	1275	119
10. Moyer and Jones[13]: H = 1194 A/m					
1232°C, H_2	7350	0.97	0.77	1600	143

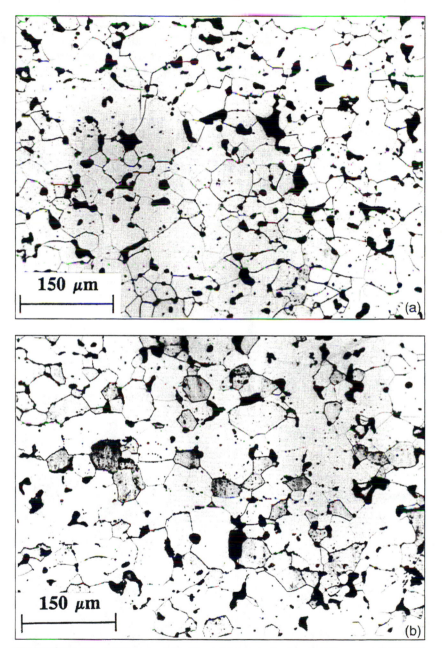

Figure 3.15: Microstructures of Ferritic stainless steels sintered in vacuum at 1260°C for 45 minutes: a) 410L and b) 434L. Ferrite matrix with well-rounded porosity. Sintered density of 7000 kg/m³.

Both C and N_2 are stabilizers of the high temperature phase, austenite. In contrast Cr is a ferrite stabilizer. The competing roles of these elements or their equivalents will determine the final microstructure. Some degree of stabilization of the high temperature austenite phase is needed before it can be turned into martensite, during the cooling cycle. The rate of cooling and the specific combination of elements will determine the amount of martensite that is formed. If the rate of cooling is too slow, bainitic structures are obtained instead of the martensitic phase.

By sintering at a high temperature, the degree of solubility of N_2 decreases so that the austenite stabilization is less. On cooling, small amounts of martensite/bainite may form as shown in Figures 3.16a and c. If these materials are sintered at a lower temperature the amount of N_2 absorbed is much greater. When this product is cooled to room temperature, almost 100% martensite is obtained (Figures 3.16b and d).

In addition to the high nitrogen absorption at low sintering temperatures, the added complication of oxide stability becomes important. As in the case of silicon steels, ferritic steels have oxides that are very stable. In the case

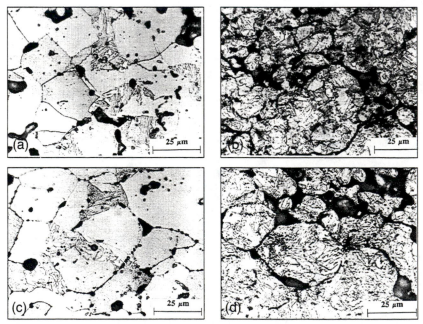

Figure 3.16: Micrographs of ferritic stainless steels showing martensitic/bainitic phases due to sintering in dissociated ammonia:

a) 410L sintered at 1288°C.   c) 434L sintered at 1288°C.
b) 410L sintered at 1121°C.   d) 434L sintered at 1121°C.

of stainless steels, the major element responsible for these oxides is chromium. A small amount of silicon is also added to the P/M stainless steels, by the powder producers, to stabilize their product. This Si can also create the same difficulties, with regard to oxide formation. An additional problem is that Cr is very sensitive to the presence of C and N. Consequently, it is recommended that stainless steels be sintered in clean, dry hydrogen or vacuum at high temperatures[11]. The higher temperatures aid the dissociation of the oxides as well as minimizing the absorption of nitrogen[15].

3.7 Additional Alloys

Table 3.9 shows the magnetic data for a number of other alloys that are used in soft magnetic applications.

Permendur is a trade name for the Fe-50% Co alloy. As mentioned earlier, this material has the highest flux density[2] of any wrought material. Takada, et al.[16] report that the alloy is often produced by the lost wax process, but can be double pressed and doubled sintered to about 98% of the full density. Such a product has excellent flux density capability as shown in Table 3.9. The oxygen level should be kept lower than 200 ppm to provide sufficient ductility for the repressing of the product.

Moly permalloy has an exceptionally low coercive field, but the flux density capability is significantly limited. This material would also benefit from processes utilizing double pressing and double sintering.

Table 3.9 Magnetic Data on Special P/M Alloys

SOURCE		PROPERTY			
	DENSITY	B_m	B_r	μ_r	H_c
	(kg/m^3)	(T)	(T)		(A/m)
Fe-81%Ni-2%Mo; Permaloy					
1. Lall and Baum[1]: H = 1990 A/m					
1260°C, DA, 45 min.	7800	0.72	0.48	77000	85
Fe-50%Co-2%V; Permendur					
2. Lall and Baum[1]: H = 1990 A/m					
1260°C, DA, 45 min.	7200	1.24	0.73		175
	7400	1.47	1.29	4700	119

Table 3.9 (Cont.) Magnetic Data on Special P/M Alloys

SOURCE	PROPERTY				
	DENSITY	B_m	B_r	μ_r	H_c
	(kg/m^3)	(T)	(T)		(A/m)
3. Takada et al.[16]: H = 7958 A/m					
1200°C, H_2, 3 hrs.	8000	2.3	—	5700	96
Fe-0.8% P					
4. Lall and Baum[1]: H = 1990 A/m					
1260°C, DA	7000	1.27	1.08		118
45 min.	7200	1.32	1.13		119
	7400	1.42	1.15		69
(Resistivity 0.32, 0.30, 0.28 micro-ohm·m at 7000, 7200, and 7400 kg/m³, respectively).					
5. Gagne et al.[23]: H = 1194 A/m					
1120°C	7400	1.29	1.25	5200	95
1260°C	7500	1.35	1.32	6000	80
6. Moyer[24]: H = 1194 A/m					
1121°C, $H_2 + N_2$	7100	1.23	1.20	4100	115
Shot-peened	7100	0.98	0.83	2100	143
7. Mossner[25]					
1121°C	7200	1.28	1.18	5300	95
1288°C	7400	1.42	1.30	9100	29

The Fe - 0.8% P alloy is often used as an improved material beyond the popular Fe - 0.45% P alloy. However, the former material exhibits greater shrinkage and distortion because of the higher level of liquid phases at a given temperature. Thus the improved soft magnetic properties are at the expense of dimensional control. Repressing of the sintered product is therefore required.

Chung, et al.[17] were able to process Fe-6.5% Si by P/M. They also described the effect of adding Al, P, Sb, B, or C on the soft magnetic properties of the alloy. Carbon was found to be the most effective element in aiding reduction of silicon oxides and therefore, the sintering step. However, the C level must be maintained well below 0.5%, since excessive amounts will remain in the product and deteriorate the intrinsic magnetic properties. Metallographic evidence demonstrated the dramatic increase in grain size and the reduction in the number of pores. The thought is that the added carbon reacts with the oxygen and minimizes the formation of silicon oxides.

3.8 Magnetic Properties of MIM Components

Table 3.10 shows some magnetic data for Fe-2% Ni, Fe-50% Ni, Fe-3% Si, Fe-6% Si and 430L processed by MIM. Densities are generally greater than the conventional P/M-processed materials discussed earlier. Naturally, there is an accompanying improvement in magnetic properties. The most significant change is in H_c and B_m for Fe-3% Si. Maximum induction is seen to double from one group of Fe-3% Si toroids to another (Table 3.10) and is thought to be due to a reduction in contamination, as discussed earlier. In MIM, as practiced by Remington Arms Company (DuPont), the interplay of C, O_2, and N_2 is very important. Similar to other binder systems, the one in use here is organic with many hydrocarbon chains. These chains are broken during the thermal processing step and the hydrogen and carbon elements are removed by oxidation. This oxidation must be carried out fairly well during the thermal binder removal step in air. Of course this leads to both O_2 and N_2 absorption. During the sintering step, the hydrogen-based atmosphere and the high temperatures serve to reverse this effect and deplete the product of O_2 and N_2.

In MIM, the balancing of C, O_2, and N_2 levels is very tricky. On the one hand, insufficient oxidation during binder removal can lead to high C levels. On the other hand, too much oxidation will result in high O_2 and N_2 levels. Typical impurity levels being achieved by MIM technology on a production basis are currently 0.006% C, 0.0042% O_2, and 0.0024% N_2, which are exceptionally good.

Table 3.10 Magnetic Data on Metal Injection Molded Alloys

SOURCE	PROPERTY				
	DENSITY	B_m	B_r	μ_r	H_c
	(kg/m^3)	(T)	(T)		(A/m)
After Lall and Baum[1]					
Fe-2% Ni, 1316°C	7670 97%	1.51	1.29		82
Fe-50% Ni, 1316°C	7660 94%	1.27	0.42		16
Fe-3% Si, 1316°C	7550 98%	1.50	1.21		45
Then, H$_2$ Annealed	7550 98%	1.47	1.14		47
Fe-3% Si, 1232°C	7540 98%	1.44	0.62		60
Fe-6% Si, 1316°C	7540 99%	1.32	1.12		46
Then, H$_2$ Annealed	7410 98%	1.37	1.22		46
After Baum[24]					
Fe, 1371°C, DA	7600 97%	1.53	1.37		294
Fe, 1371°C, DA	7550 96%	1.55	1.34		183
Fe-3% Si, 1371°C, DA	7550 98%	1.45	1.07		57
Fe-3% Si, 1371°C, DA	7550 98%	1.50	1.21		51
Fe-6% Si, 1371°C, DA	7420 99%	1.37	1.21		40
Fe-50% Ni, 1371°C, DA	7660 93%	1.27	0.42		16
430L SS, 1371°C, VAC	7400 95%	1.15	0.54		202
After Mossner[25]					
Fe-3% Si		1.48	0.88		62
		1.46	1.10		56
Fe-50% Ni		1.22	0.53		16
		1.04	0.64		19
430L SS		0.85	0.22		119

3.9 Discussion of Factors Affecting Performance

3.9.1 *Resistivity*

High performance soft magnetic materials should magnetize and demagnetize easily; hence, H_c should be small, while μ_m and B_m should be as large as possible. Furthermore, if the part is going to be used for pulsed d.c. or a.c. applications, the electrical resistivity (ρ) should be as high as possible, in order to minimize eddy current losses and the resultant heat build up as related by Equation [1.17] and discussed in sub-section 1.5.

Each of the P/M materials discussed earlier, have different values for the electrical resistivity, so that each will respond with a different energy loss. As shown in the tables, the electrical resistivity increases in the following order for the most popular P/M magnetic materials: a) Fe, b) Fe-0.45% P, c) Fe-3% Si, and d) Fe-50% Ni. The same sequence, therefore, indicates the order of increasing suitability of the material for a.c.-type applications. In general, this is also the sequence in which the width of the hysteresis curve decreases, suggesting the additional benefit of lower hysteresis losses. The resistivity of P/M parts depends on the material composition and its density. These are discussed in Section 4.4, together with the effect on the function of printer pole pieces.

3.9.2 *Process Temperature and Time*

Throughout this sub-section, the effect of sintering temperature on the magnetic properties has been highlighted. Experience suggests that Fe-Si and Fe-Ni should be sintered at or above 1316°C (2400°F) for close to an hour. The former alloy system requires this because of the need to reduce surface oxides, while the stainless steels need it from a diffusion rate standpoint. While improvements are realized by sintering at higher temperatures, the vast majority of parts producers almost exclusively use 1121°C (2050°F) for 30 minutes or less.

Products sintered at 1121°C (2050°F), naturally, will not have the properties derived from test rings processed at the optimum conditions. While repressing and annealing may lead to similar densities, the inferior "cleanliness" or integrity of the microstructure will give less than optimum properties. The grain growth exhibited at elevated temperatures and the coalescence of pores are important phenomena for the production of superior soft magnetic parts. This required temperature will vary from one material to another; a 1121°C (2050°F) for Fe-0.45% P is, perhaps, equivalent to a 1260°C (2300°F) sinter for Fe-3% Si. The effect of time is also fairly significant, leading to excellent soft magnetic properties[8,9] by the use of 2 to 24 hours of sintering. Such times are, however, not economical for producing and selling parts in highly competitive business markets. The properties of parts are therefore a compromise between the desire to obtain better properties and the need to minimize manufacturing costs.

Another difficulty is that a final stress anneal is sometimes eliminated by some parts producers in order to further reduce costs. For optimum properties and performance, a final thermal operation should be included for parts that have been through any secondary operation that results in deformation damage.

3.9.3 *Impurities*

Over the years, a significant improvement in the purity and quality of commercially available metal powders has been a major factor in the development of better soft magnetic materials processed by P/M. Another factor has been the improving quality of clean, dry atmospheres. The soft magnetic performance of the materials under discussion is most sensitive to three elements: C (from the raw metal powder, poor lubricant removal, or as a contaminant), O_2 (also from the raw powder, or due to poor furnace atmosphere control) and N_2 (from the powder, the sintering atmosphere or from influx of air into the furnace). While the individual levels of these elements are important, it is suggested that their combined levels with each other are also critical to soft magnetic performance.

In 1974, Baum[18] showed that both C and N have a very significant effect on the coercive field of Fe-3% Si, with N_2 being the most serious impurity (Figure 3.17). A later study[19] by Moyer and McDermott looked at the magnetic properties of Fe and Fe-P alloys in hydrogen-nitrogen sintering atmospheres (1121°C [2050°F], for 30 minutes). With C being maintained at less than 0.01 wt. %, they showed that the best atmosphere was one of pure H_2, with 100% dissociated ammonia (75% H_2 = 25% N_2) being the second choice. They showed that with more than about 80% N_2, the O_2 and N_2 levels increase beyond about 0.03% and 0.007%, respectively. At these levels and above, the magnetic properties degrade rapidly. Again, N_2 was the impurity to which the coercive field and maximum permeability were most sensitive. Furthermore, it was shown that the poor practice of sintering carbon-containing parts with magnetic iron parts at 1121°C (2050°F) in DA, can result in a pickup of 0.03% C. This leads to a significant decrease in μ_m, B_m, and an increase in H_c.

A recent study[20] on the absorption of C-O-N in soft magnetic P/M materials showed that minimal nitrogen (less than 50 ppm) is absorbed in Fe, Fe-0.45% P, Fe-3% Si and Fe-50% Ni when sintered in 100% DA or

80

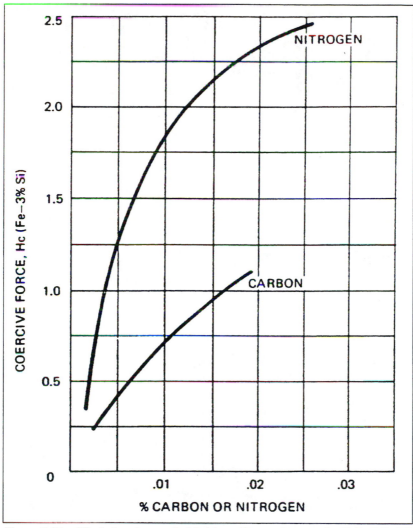

Figure 3.17: Effect of carbon and nickel impurities on the coercive field of Fe-3% Si[18].

mixture of 75% H_2 + 25% N_2. Similar results were obtained by vacuum sintering with a nitrogen backfill and nitrogen gas quench. Carbon and oxygen levels were in the range of a few hundred ppm.

Such was not the case for ferritic stainless steels sintered in these same atmospheres. While carbon was maintained at a few hundred ppm, both oxygen and nitrogen absorption levels were an order of magnitude higher. No doubt, this is due in part to the strong affinity of Cr for both of these elements. The level of nitrogen absorbed decreases with increasing temperature above 1120°C, similiar to the austenitic stainless steels[15], and is demonstrated by the lower amounts of martensite observed in the microstructures of section 3.6. Nitrogen absorption was minimal in the vacuum sintered product.

It is suggested that the impurity levels should to be controlled to the following levels when considering high performance soft magnetic applications:

Carbon   -   0.01% max

Oxygen   -   0.02% max

Nitrogen   -   0.01% max

The preferred upper limits are about half of those indicated above. These impurities lead to the formation of precipitates or inclusions which become a hindrance to the motion of magnetic domain walls (see section 1.4). In high performance soft magnetic applications, this is perhaps the most dominant phenomenon throughout the cycle. Anything that prevents domain wall pinning will improve soft magnetic properties. The formation of these precipitates over a period of time causes a decrease in magnetic properties[19] and is known as "**magnetic aging**". The degradation of properties as a result of nitrogen absorption during sintering, was found to be severest when samples were aged at 100°C (212°F). It is well known that magnetic aging decreases with increasing silicon content, in the Fe-Si alloy system.

Table 3.11 shows the effect of intentional additions of certain elements on magnetic properties. By admixing 0.5% C into Fe the value of H_c is essentially tripled. If the familiar Fe-2%Cu-0.8% C blend is processed and tested, the coercive field is found be almost five times that for pure Fe. In addition, the flux density values decrease significantly. These demonstrate the serious degradation of soft magnetic properties, making the product unsuitable for demanding applications.

Table 3.11 Effect of Alloying Elements on Magnetic Properties of Iron

SOURCE		PROPERTY			
	DENSITY	B_m	B_r	μ_r	H_c
	(kg/m^3)	(T)	(T)		(A/m)
POWDER METAL:					
Pure Fe					
1. Lall and Baum[1]: H = 1990 A/m					
1260°C, DA	6800	1.14	0.96	2900	131
45 min.	7200	1.36	1.10	3700	127
	7400	1.47	1.29	4700	119
2. Lall[20]: H = 1990 A/m					
1249°C, DA, 50 min.	7000	1.15	1.08	2400	135
Fe-0.5% C					
3. Lall[20]: H = 1990 A/m					
1249°C, DA, 50 min.	7000	1.08	0.95		382
Fe-2%Cu-0.8%C					
4. Lall[20]: H = 1990 A/m					
1249°C, DA, 50 min.	7000	1.95	0.8		597

3.10 Comparison of P/M and Wrought Products

Table 3.12 is a simplified summary chart that compares the soft magnetic properties of wrought and powder metal products. Those properties of P/M components that are commonly achievable in industrial environments are listed, provided that proper processing conditions are used. Data that is only attainable under tight laboratory conditions has been specifically excluded. In the same way, the wrought material properties are also reflective of products that are made routinely; neither the best, nor the worst properties have been selected. Clearly, for a fair one-to-one comparison, only randomly oriented polycrystalline wrought products were selected. The grain-oriented (textured) materials are clearly far superior to any equivalent P/M product of the same composition.

Table 3.12 Comparison of Materials Produced by Wrought and Powder Metallurgy Techniques

	Fe	Fe-0.45% P	Fe-3% Si	Fe-50% Ni
MAXIMUM INDUCTION, B_m (T)				
P/M	1.4	1.4	1.3	1.2
WROUGHT	1.7	-	1.6	1.5
RESIDUAL INDUCTION, B_r (T)				
P/M	1.2	1.2	1.1	0.9
WROUGHT	1.3	-	0.7	0.8
COERCIVE FIELD, H_c (A/m)				
P/M	127	80	80	16
WROUGHT	80	-	64	8
MAXIMUM PERMEABILITY, μ_r				
P/M	4,000	5,000	5,000	20,000
WROUGHT	5,000	-	7,000	70,000
ELECTRICAL RESISTIVITY (micro-ohm·m)				
P/M	0.12	0.20	0.55	0.70
WROUGHT	0.10	-	0.50	0.50
CORE LOSS (watts/kg)				
P/M	24	16	5	-
WROUGHT	12	-	4	-
DENSITY of P/M				
kg/m^3	7200	7200	7200	7400
% OF WROUGHT	92	92	94	90

With the additional porosity in P/M components, the flux density as indicated by B_m and B_r is naturally lower than the equivalent wrought compositions. The effect is more than a linear relationship since the internal porosity tends to set up demagnetizing fields[4]. Utilization of double pressing and sintering, together with high temperature vacuum sintering, brings the wrought and P/M flux densities closer together.

In terms of coercive field, the P/M products have slightly higher values since the pores tend to act as pinning points for domain wall motion. Furthermore, the pores inhibit grain growth. Consequently, the smaller grains in P/M tend to increase H_c since the grain boundaries also are a hindrance to the motion of Bloch walls. On the plus side, the materials used in P/M are considerably cleaner, and can be maintained this way with a little care during processing.

On another positive note, the electrical resistivity of the P/M materials is slightly higher, because of the added porosity in P/M. The difference, however, is not sufficient to have a major impact on core losses. Furthermore, it is far more practical and economical to form laminated structures from wrought sheets than P/M. The core losses in a device using wrought laminated sheets will be considerably lower than if the device was made from a solid P/M part.

On the other hand, the P/M process has the flexibility to economically custom-blend small lots of specialty alloys. All one has to do is to select the right powders, weigh them out and put them in a production blender. A variety of sizes are available, but normal production quantities are in the range of 227 to 13,605 kg (500 to 30,000 lbs).

Powder metallurgy also has the advantage of being able to process very difficult materials. For example, Fe-P alloys cannot be made by conventional wrought metallurgy techniques because of the "hot shortness" failure during the elevated temperature rolling step. With P/M processing, this is not a problem and, in fact, the transient liquid phase assists densification during sintering. In the Fe-Si system, the Si level has to be kept low in order to minimize the increase in hardness and embrittlement, which present serious dificulties in the rolling of wrought product. P/M also has similar problems during the compaction stage, but slightly higher levels of Si can be processed. Furthermore, the P/M process makes use of the transient liquid phase in this alloy system to aid the consolidation of the product.

The metal injection molding process provides an even greater degree of flexibility in its ability to work with materials that traditionally would have been impractical. For example, almost any level of Si can be selected and the components produced to completion. The Fe-6% Si alloys, for which the data are shown in Table 3.10, presented no difficulties during the MIM processing to near-net shapes.

REFERENCES

1. C. Lall and L. W. Baum, Jr., "High Performance Soft Magnetic Components By Powder Metallurgy and Metal Injection Molding", *Modern Developments in Powder Metallurgy*, Vol. 18, compiled by P. U. Gummeson and D. A. Gustafson, Metal Powder Industries Federation, Princeton, NJ, 1988, pp. 363-389.

2. Metals Handbook, 8th Edition ASM International, Materials Park, OH, 1984, Vol. 1, pp. 785-797.

3. K. H. Moyer, M. J. McDermott, M. J. Topolski and D. F. Kearney, "Magnetic Properties of Iron Alloys", *Magnetic and Electrical P/M Technology and Applications,* P/M Seminar, Metal Powder Industries Federation, Princeton, NJ, 1980, pp. 37.

4. E. Adler, G.W. Reppel, W. Rodewald, H. Warlimont, "Matching P/M and the Physics of Magnetic Materials", *Int. J. Powder Metall.,* 1989, Vol. 25, No. 4, pp. 319-335.

5. K. Moyer, and J. Ryan,"Iron-phosphorus Alloys for Magnetic Applications-Do We Know Everything About These Alloys?", *Progress in P/M,* 1987, Vol. 43, pp. 323-336.

6. F. V. Lenel, U. S. Patent No. 2,226,520, (1940).

7. P. Lindskog, J. Tengzelius, and S. A. Kvist, "Phosphorus As An Alloying Element in Ferrous P/M", *Modern Developments in Powder Metallurgy,* Vol. 10, Metal Powder Industries Federation, Princeton, NJ, 1984, pp. 133-151.

8. B. Weglinski, and J. Kaczmar, "Effect of Fe_3P Addition on Magnetic Properties and Structure of Sintered Iron", *Powder Metallurgy,* The Metals Society, London, England, 1980, Vol. 23, No. 4, pp. 210-216.

9. J. Kaczmar and B. Weglinski, "Influence of Processing Parameters on Magnetic Properties of Fe-0.8% P Sintered Materials", *Powder Metallurgy,* The Metals Society, London, England, 1980, Vol. 27, No. 1, pp. 9-13.

10. L. W. Baum, Jr., "Theoretical and Practical Considerations for the Production of P/M Parts", *Hoeganaes P/M Technical Conference,* Philadelpia, PA, 1978.

11. C. Lall, "Fundamentals of High Temperature Sintering: Application to Stainless Steels and Soft Magnetic Alloys", *Int. J. Powder Metall.,* 1991, Vol. 27, No. 4, pp. 315-329.

12. C. Lall, Internal Reports, 1982-1984, Remington Arms Company (DuPont).

13. K. Moyer, and W. Jones, "Stainless Steels for Improved Corrosion Resistance", *Advances in Powder Metallurgy-1991,* Vol. 4, compiled by L. F. Pease and R. J. Sansoucy, Metal Powder Industries Federation, Princeton, NJ, pp. 145-158.

14. M. Svilar and H. D. Ambs, "P/M Martensitic Stainless Steels: Processing and Properties", *Advances in Powder Metallurgy-1990,* Vol. 2, compiled by T. G. Gasbarre and W. F. Jandeska, Metal Powder Industries Federation, Princeton, NJ, pp. 259-272.

15. H. S. Nayer, R.M. German and W. R. Johnson, "The Effect of Sintering on the Corrosion Resistance of 316L Stainless Steel", *Progress in Powder Metallurgy,* Vol. 37, compiled by J. M. Capus and D. L. Dyke, Metal Powder Industries Federation, Princeton, NJ, 1982, p. 225.

16. T. Takada, K. Asaka and S. Tomita, "Magnetic Properties of Sintered Permendur", *Advances in Powder Metallurgy-1990,* Vol. 2, compiled by T. G. Gasbarre and W. F. Jandeska, Metal Powder Industries Federation, Princeton, NJ, pp. 421-429.

17. H. S. Chung, T. H. Yim, Y. J. Kim, W. K. Kang and Y. H. Jeong, "The Effect of Sintering and Secondary Elements on the Soft Magnetic Properties of P/M Fe-6.5% Si Steel", *Advances in Powder Metallurgy-1991,* Vol. 5, Metal

Powder Industries Federation, Princeton, NJ, pp. 169-178.

18. L. W. Baum, *Precision Metal,* March, 1974.

19. K. H. Moyer, and M. J. McDermott, "The Effect of Sintering in Hydrogen-Nitrogen Atmospheres and Subsequent Aging on the Magnetic Properties of Iron", *Powder Metall. Int.,* 1983, Vol. 15, No. 3, Verlag Schmid, Freiburg, Germany.

20. C. Lall, "The Effect of Sintering Temperature and Atmosphere on the Soft Magnetic Properties of P/M Materials", *Advances in Powder Metallurgy-1992,* Vol. 3, Metal Powder Industries Federation, Princeton, NJ, 1992.

21. M. J. McDermott, 1992, Hoeganaes Corp., Riverton, NJ, private communication.

22. M. Gagne, J.P. Poirier, and Y. Trudel, "Designing a Steel Powder for Soft Magnetic Applications", *Advances in Powder Metallurgy-1990,* Vol. 2, compiled by T. G. Gasbarre and W. F. Jandeska, Metal Powder Industries Federation, Princeton, NJ, pp. 407-420.

23. K. Moyer, "Secondary Operations and Effects on Magnetic and Tensile Properties", *Advances in Powder Metallurgy-1990,* Vol. 1, compiled by T. G. Gasbarre and W. F. Jandeska, Metal Powder Industries Federation, Princeton, NJ, pp. 521-528.

24. L. W. Baum, 1992, Remington Arms Co., Ilion, NY, private communication.

25. W. Mossner, 1992, SSI, Jamestown, WI, private communication.

26. K. H. Moyer and J. B. Ryan, "Emerging P/M Alloys for Magnetic Applications", *Modern Developments in Powder Metallurgy,* Vol. 18, compiled by P. U. Gummeson and D. A. Gustafson, Metal Powder Industries Federation, Princeton, NJ, 1988, pp. 757-3772.

27. "Carpenter Soft Magnetic Alloys", Carpenter Technology Bulletin 2-91/1M, Carpenter Steel Division, Reading, PA, 1991.

Chapter **IV.**

EXAMPLES OF
SOFT MAGNETIC
DEVICES

In this concluding chapter, a number of parts made by powder metallurgy and metal injection molding are described. The characteristics of the components made by these forming techniques are discussed, as well as the effect of the component properties on the soft magnetic function of the application.

4.1 Introduction

The remarkable advances in the fields of materials science, electronics and magnetism have lead to a myriad of electronic monitors and controls for industrial and home use. Perhaps the most important of these electronic instruments is the computer, which in the latter half of the twentieth century has revolutionized the way society functions. The profusion of personal computers (PC's) portends of changes that will affect forever the way business and personal affairs may be conducted in the future.

In every kind of environment, computers are used with magnetic media for data storage onto floppy or hard discs. The linear voice coil motor and positioning devices for the read-write heads depend on magnetism. Dot-matrix and line printers rely on magnetic materials for printing characters. A variety of printers, FAX and copying machines rely on magnetism to transport the paper. Financial institutions depend on computers for fund transfers and electronic communications. Neither national nor international commerce could function at todays pace without these facilities.

Computers and other electronic processing units are often used to monitor and/or control external functions or operations. In the majority of cases, where some mechanical motion is involved, the interfacing mechanism will utilize some form of electromagnet device. One simple example is a solenoid with a movable soft magnetic component. A constant force is applied in one direction by a spring. This force is overcome by energizing the solenoid and causing motion of the soft magnetic component.

In industry, magnetism is used in motors, generators, power tools, metering devices, magnetic separators, pumps, proximity switches, etc. In the automobile, magnetism is prevalent in fuel pumps, fuel injectors, starter motors, seat positioners, door locks, window lifts, anti-lock brake systems, positive traction control, headlamp and rear-view mirror positioners, etc.

At home, a variety of kitchen appliances from the lowly electric can opener to the microwave oven depend on magnetism to function. Additional examples are dishwashers, clothes washers, food blenders, trash compactors, etc. In other parts of the house, electric shavers, clocks, entertainment systems (video cassette recorders-televisions-radios), and many children's toys depend on this phenomenon.

The medical profession uses magnetic assemblies in fluid delivery systems, magnetic resonance imaging (MRI), computer aided tomography (CAT) scanners, patient monitoring instruments employing cathode ray tubes (CRT), etc. Mass spectrometers and machines that focus positive ions and electrons are additional examples where the science of magnetism plays a key role.

Many of the examples cited utilize permanent magnets and soft magnets or a combination of both. Sometimes no magnet is used at all, as in the case of the magnetic field generated by an electric current in a wire. Most often, however, a soft magnetic material is used in conjunction with a coil or permanent magnet to intensify and direct the magnetic lines of flux.

In this concluding chapter, examples of soft magnetic components manufactured by powder metallurgy (P/M) and metal injection molding (MIM) are described. The selected examples have been chosen to illustrate the inter-relationship of material properties, processing, and component function. Furthermore, guidance is given as to which magnetic attributes need to be altered to optimize the system performance or efficiency. Figures 4.1 and 4.2 show examples of shapes of parts that can be made by powder metallurgy and metal injection molding[1], respectively. Figure 4.1 shows a couple of motor stators, a rotor, an E-core, pole-piece armatures, and pole pieces for line and dot matrix printers. The MIM examples in Figure 4.2 include a dot matrix head and components for the aerospace and automotive industries. The examples shown in these figures include the four basic materials; i.e., Fe, Fe-0.45% P, Fe-3% Si and Fe-50% Ni.

4.2 Solenoids and Fluid Flow Control Devices

Figure 4.3 is a schematic of a solenoid with a metal housing and a moving armature or "piston" which is held in position by a spring. As the

Figure 4.1: Examples of some magnetic components made by powder metallurgy[1].

Figure 4.2: Examples of some magnetic components made by metal injection molding[1].

solenoid is energized, a magnetic flux is generated in the surrounding materials. Since the lines of force run across the air gap (perpendicular to the faces), the movable "piston" will be pulled axially, overcoming the spring force. When de-energized, the system loses its magnetic field and the spring pushes the armature back into the original starting position. Such a device can be used to perform a number of short-stroke functions. This includes opening and closing valves and switches.

A slight modification of the schematic in Figure 4.3 produces a device that can control fluid flow in an axial direction, through the electro-magnetic assembly. Figure 4.4 shows a version in which the armature is restrained away from the opening by a spring. When the coil is energized the "piston" moves to the right closing the hole and cutting off fluid flow. Such an assembly is used to control hydraulic brake fluid flow in electronic anti-lock brake systems. The electro-magnetic valve is shown in the normally open position, but can also be set as normally closed.

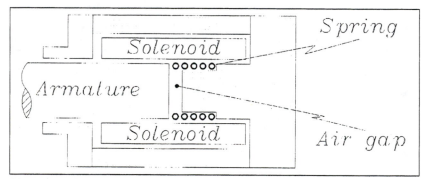

Figure 4.3: Schematic of a solenoid unit.

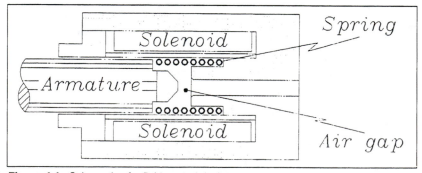

Figure 4.4: Schematic of a fluid control device.

A similar device may be used for control of fuel flow into the engine intake manifold. One fuel injector system uses a separate main fuel pumping source and the "injectors" simply act as valves, controlling the amount of fuel to be delivered to each cylinder. Another system relies on the piston or armature to actually "pump" or "inject" the fuel into the engine cylinder. A pulsating d.c. signal applied to the solenoid causes the armature to oscillate back and forth, resulting in the fuel being pumped into the opening on the right (Figure 4.4).

In low cycle a.c. or pulsating d.c. applications, such as the above, the primary aim is to have a high value of B_m which enables the material to carry the largest amount of flux across the lowest cross-sectional area. Eddy current losses are not a significant issue. A high B_m value is obtained by maximizing the P/M part density. If a fast response is required, the value of μ_m should be high and H_c should be low.

Pure Fe or Fe-0.45% P can be used in most applications. For applications operating at higher frequencies, the actuator may function better if it is made from Fe-3% Si. This material will provide a better dynamic response and impact resistance. Constant impacting at the air gap can cause localized deformation which will be minimized by the higher hardness values of the Fe-3% Si alloy.

Sometimes, magnetic performance is not the solution to a problem. In the case of the fuel injector, described above, one customer was having difficulty in the assembly of the housing (the outside body of Fig. 4.4). Although magnetically the housing served its role, the edges of the "cup" were showing cracks when the housing was crimped over an end-plate. In order to improve the ductility, several options were considered: increasing the sintering temperature and time, pressing to a higher density, using H_2 or vacuum instead of cracked ammonia (DA) or methane gas. A higher temperature sinter in pure hydrogen resolved the ductility concerns.

4.3 Anti-lock Brake Systems and Positive Traction Control

Anti-lock Brake Systems (ABS) were developed in the 1970's and 1980's with the purpose of electronically overcoming the driver's tendency to overbrake in emergency situations. Under wet, icy, oily, or other slippery road conditions, the wheels can be locked up and skid, if too much braking pressure is applied. In a similar manner, Positive Traction Control (PTC) minimises the skidding during excessive acceleration by modifying fuel intake. The operating principles are almost identical for ABS and PTC devices.

An anti-lock brake system consists of the sensor ring, the detecting coil

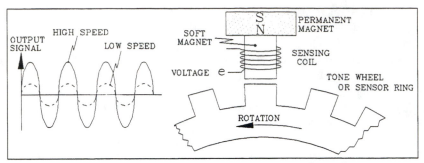

Figure 4.5: Schematic, displaying components of an Anti-lock Brake System.

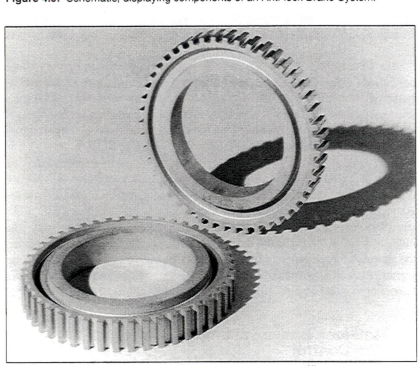

Figure 4.6: Example of an Anti-lock Brake System sensor ring[4]. Courtesy of Carbon City Products, St. Mary's, Pennsylvania, U.S.A.

("sensor"), the hydraulic brake system and the electronic control unit. P/M components are used for the sensor ring and the sensor (Figure 4.5), as well as the hydraulic fluid control devices mentioned in the previous section. One configuration of an ABS sensor ring that has been used in production is shown in Figure 4.6. Sensor rings may be positioned on the wheel axle or in the transmission train.

94

A permanent magnet attached to a soft magnetic pole piece serves as a source of magnetic flux. As the sensor ring, also called a "tone wheel" rotates, the magnetic flux, Φ, is changed, generating a voltage in the coil of N turns, according to Faraday's Law, which was given above in Equation [1.21].

$$e = -N \frac{d\Phi}{dt} \qquad [4.1]$$

The size of the magnetic flux is related to the air gap, which should be a) as small as possible and b) consistent. The rate at which the flux is changed will depend on the number of teeth (n) on the sensor ring and the rate of rotation. If the rate of rotation, w, is expressed as revolutions per minute, the frequency (f) is given by:

$$f = \frac{w}{60} n \qquad [4.2]$$

The frequency, f, is proportional to the road wheel speed and, therefore, the vehicle speed, v:

$$f = c\, v \qquad [4.3]$$

Here, the constant c is vehicle-specific, and dependent on such things as the road wheel size, location of the sensor ring, etc.

Pure Fe and Fe-0.45% P have been used for both the sensor ring and the pole piece in the sensor. The Fe-0.45% P material is preferred because of the ability to obtain higher densities as well as an improved maximum permeability. An increase in density, implies that more of the magnetic flux, Φ (supplied by the permanent magnet), is transmitted by the pole piece. A high value of μ_m means that the material will respond more readily to the changes in the flux at the higher frequencies and therefore, higher vehicle velocities. These comments apply to both the sensor ring and the sensor pole piece.

The sensor ring has also been manufactured by P/M from several 400-series stainless steels (e.g., AISI grades 410L, 434L, 454L, etc.), for improved corrision resistance. As shown in Table 3.8, the soft magnetic properties of these materials are inferior to the other P/M materials. This reinforces the thought that perhaps the most important aspect of the sensor ring is that it should be a high density ferromagnetic material. A high value for μ_m appears to be unhelpful since the voltage generated at high frequencies is reasonably high. The greater problem with ABS control and feedback is at low vehicle speeds, where the rate of flux change is low and the induced voltage is, therefore, also low. If the signal-to-ratio becomes too small (because of a low wheel speed, or variations in the air gap or tooth density), the electronic control module cannot differentiate between random noise and a genuine signal from the sensor ring. Such situations can make it difficult to determine whether the sensor ring is static or actually rotating.

The ferritic stainless steels have to be processed carefully, or the soft magnetic properties can be unacceptably poor[2,3]. Stainless steels are particularly sensitive to contamination by C, O_2, and N_2 because of their affinity for Cr. The effect of nitrogen and carbon in stabilizing the austenite phase in the ferritic stainless steels, thereby introducing martensite during cooling, was discussed in the previous chapter. For this reason, sintering of stainless steels in atmospheres containing nitrogen should be avoided.

Stainless steels should be processed at high temperatures and in pure hydrogen or vacuum to obtain the optimum soft magnetic properties. An additional benefit with this optimized processing is that the ductility is increased significantly. This is an important property for some manufacturers since the sensor ring is assembled by press-fitting onto the wheel axle. Early attempts to convert from wrought (machined) sensor rings to P/M experienced such failures because of poor ductility[2]. As in the case of the fuel injector housing discussed in the previous sub-section, this illustrates a case where another property is equally important as the magnetic properties, in fulfilling the customer's needs.

While many sensor rings are made from ferritic stainless steels, the trend appears to be towards the use of pure Fe or Fe-0.45% P, which are covered by a corrosion-resistant coating. Even with the additional operation, the lower cost raw powders and ease of processing favor the use of iron and iron-phosphorus materials.

4.4   Relays and Printer Actuator Mechanisms

In an E-core (bottom center of Figure 4.1), a coil is wound onto one or more sections of the frame and energized. The current flowing through the coil induces a magnetic field in the core which tends to pull a soft magnetic material towards it, across a small air gap (Figure 4.7). The closure piece, called the armature or "clapper", is usually restrained away from the E-core by a spring. The strength of the induced field must, therefore, be strong enough to overcome this spring force.

In most cases, the core is activated under d.c. or fluctuating d.c. modes. Therefore, the primary requisite of the soft magnetic material is to carry the maximum flux density (high B_m) at a minimum energy (low H_c and high μ_m). This is similar to the devices discussed in sub-section 4.2. The materials often selected, for the same reasons, are pure iron or iron-

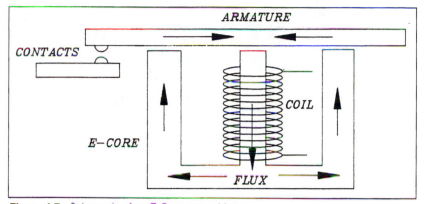

Figure 4.7: Schematic of an E-Core assembly.

phosphorus. Eddy current losses are usually not important. The main requirement is for the material to carry as much flux as possible within the design configuration. The d.c. signal sent to the coil can be used to activate relays or switches.

A number of printer actuator mechanisms were developed in the seventies and eighties, based on similar principles. These pole pieces (several versions shown in Figure 4.1) are used to activate dot matrix and line printers. The latter print a whole character at a time, while the matrix printers form a character from a series of dots.

Figure 4.8 is a schematic of a fixture that was used to test pole pieces intended for dot matrix printers. In the printer head, the pole pieces are arranged in a circle, with the tungsten wire assembly towards the center. The tungsten wire is threaded through several guides to form a vertical row of dots at the center of the printer head assembly. At least 9 dots are used in the high resolution printers. In operation, the printer head scans horizotally across the page as the appropriate pole pieces are "fired" to build up portions of an alphanumeric character. An advantage of dot matrix printers over line printers is the ability to to produce a graphical output. Both types can make multiple carbon copies, whereas an ink-jet printer cannot.

When the coil in Figure 4.8 is energized, the armature ("clapper") is pulled towards the pole piece and pushes down on the end of the tungsten wire. The force is transmitted through the wire to the load cell, which would be the ink ribbon in an actual printer. A distance of 5.1×10^{-4} m (0.020 in) was preset between the load cell and the tip of the wire.

To conduct the test, a selected pole piece was loaded into the fixture and a single square wave-form applied to the coil. A digital storage oscilloscope was used to capture this input wave-form, as well as the buildup and decay

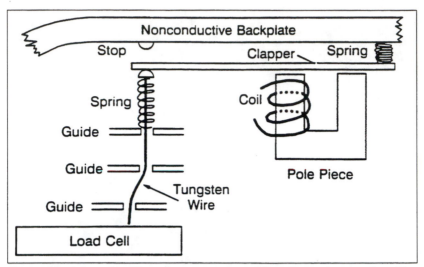

Figure 4.8: Schematic of fixture for measurement of flight time and current decay[1].

of the coil current, and the load cell output. This information was subsequently transferred electronically to an X-Y recorder.

Examples of two such outputs are shown in Figure 4.9 for pole pieces made from Fe-3% Si. The pole piece sintered at the higher temperature yielded a better performance, in that a greater impact force and a faster build-up and decay were observed. In addition, the time interval ("flight time") between the beginning of the input signal and the first detection of an impact at the load cell was shorter for the high temperature sintered product. This kind of testing can also be used to make comparisons between different materials and related to experiences in the field.

In one case history, Fe-3% Si was replaced with Fe-45% P on line printers at a customer's test facility. This resulted in "streaky" printing and poor carbon copies. A lower electrical resistivity implies a lower reluctance, which leads to a slower magnetic decay and the impact character staying too long in contact with the ink ribbon. Also, the low impact load meant that the last carbon copies did not come through clearly (in commercial applications, as many as 5-6 copies are made at the same time).

In addition, the temperature of the printing units rose significantly. The reason for this is quite apparent from the electrical resistivity data shown in Figure 4.10. The resistivity of Fe-3% Si is about 2.5 times that of Fe-0.45% P and 1.8 times that of Fe-0.8% P. The eddy current losses are reduced by the higher resistivity according to Equation [1.17], in sub-section 1.6. Note,

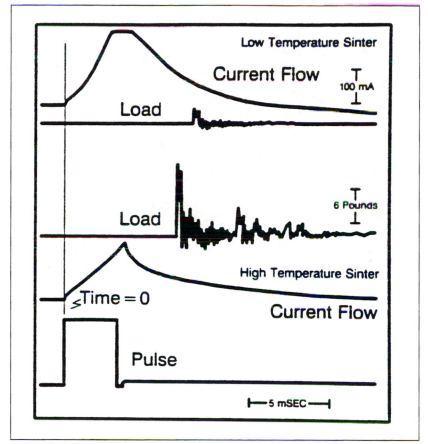

Figure 4.9: Flight time and current decay waveforms for Fe-3% Si[1].

also the much higher electrical resistivity values for Fe-6% Si and Fe-50% Ni, both of which would exhibit significantly lower eddy current losses.

From a processing viewpoint, high compaction pressures and sintering temperatures are helpful in achieving the twin goals of high densities and clean microstructures. Double pressing, with an intermediate step of a low temperature anneal, is also a good option for the attainment of these goals. Material selection can be guided by the resistivity data of Figure 4.10 and the magnetic data tables presented in chapter 3. In many cases, well-processed silicon iron or phosphorus iron can be used in printers operating at medium to high speeds.

Metal injection molded products are also being developed at this time for the computer peripherals market, but such work is largely proprietary.

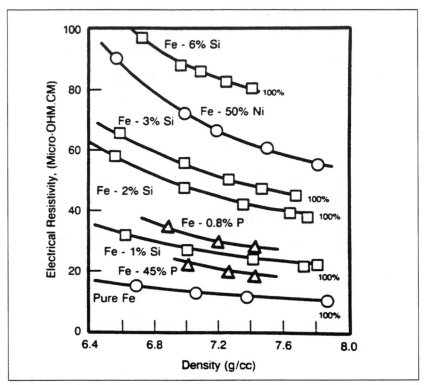

Figure 4.10: Electrical resistivity of some popular P/M alloys as a function of sintered density[1].

The photograph in Figure 4.2 shows a part at the top right in which all 9 pole pieces have been molded as one piece. This is a miniature printer head assembly of the type described earlier. The MIM technology makes it possible to fabricate such complex shapes, together with improved soft magnetic properties (see Table 3.10, sub-section 3.4), compared to P/M. See sub-section 4.6 for further discussion of MIM components.

4.5 Linear Voice Coil Motors and Disc Drive Components

The traditional well-known use for these devices has been in loudspeakers for entertainment systems, but an equally important application is as an accurate positioning tool. For example, the linear voice coil motor (LVCM) is used to precisely position the read-write heads in computer disc drives.

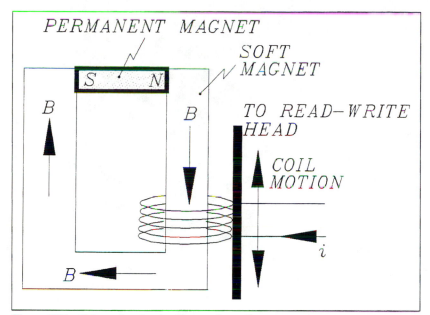

Figure 4.11: Schematic, showing linear voice coil motor components.

In the simplified schematic of Figure 4.11, a powerful permanent magnet is used to generate a magnetic field in the soft magnetic piece (core). A movable coil is positioned around one leg of the core. The position of this coil is changed by simply varying the current flowing through the coil. With precise control of the current flowing through the coil, the position of the read-write head (which is attached to the coil by a lever arrangement) can be controlled very accurately.

The most important attribute of the soft magnetic core material is flux density. A low H_c and high maximum permeability are secondary requirements. As such, pure Fe and Fe-0.45% P can be used effectively in these applications. High compaction pressures and sintering temperatures are recommended to enhance densification. Double pressing and double sintering may be the optimized approach for processing. Both large and small components are being used in disc drive systems. Their primary function is to carry magnetic flux, so that B_m is the most important property that is required. Density is therefore the highest priority. Many disc drive components are resin impregnated and nickel plated to minimize corrosion as well as maintaining a clean environment in the hardware. The materials of choice are pure iron or Fe-45% P.

Figure 4.12 shows an actual voice coil motor assembly consisting of

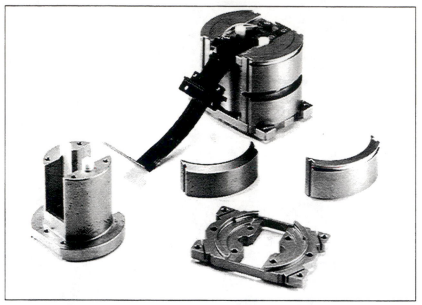

Figure 4.12: P/M components of a voice coil actuator assembly for a hard disc drive[4]. Courtesy of AMES S.A., Barcelona, Spain.

four parts made by P/M that are in production[4]. The parts are made from pure iron fabricated to a high density (greater than 7300 kg/m^3) to maximize the value B_m. This assembly of P/M parts was found to have a higher performance than one made from investment cast parts.

4.6 Examples of Parts Made by MIM

Several MIM parts are in production and a number of products are being developed for the smaller disc drives, but the proprietary nature of the work precludes detailed discussion. Some fairly complex shapes, including arcs or other curved sections may be formed. The obvious advantages are the opportunity to form complex shapes and the improved material properties compared to P/M. Pease[5,6] has given several examples of parts that have been made by MIM technology. However, only one example of a soft magnetic part, a stepper motor rotor, is shown.

Figure 4.13 shows several MIM parts[7] that are used in a solenoid assembly requiring good soft magnetic properties, high yield strength and high ductility. An Fe-Ni alloy was chosen to provide the right combination of properties. This is another example where the mechanical properties were equally important as the magnetic properties to meet the functional

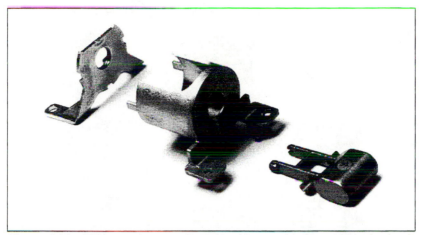

Figure 4.13: MIM components of a solenoid assembly for an electric circuit breaker[7]. Courtesy of Parmatech Corporation, Petaluma, California, U.S.A.

Figure 4.14: 24-pin and 9-pin dot matrix heads made from Fe-2% Ni by the MIM process. Courtesy of Parmatech Corporation, Petaluma, California, U.S.A.

requirements. At a density of about 7800 kg/m^3 a B_m value of 1.4 tesla was obtained. By using pure iron, the maximum induction would have been higher, but the yield strength needed in the application would not have been achieved. The MIM forming technology also enabled several parts to be combined into one, thereby reducing the number of assembly operations.

Figure 4.14 shows two printer head components for dot matrix printers manufactured by MIM. These are examples of 24-pin and 9-pin heads

Figure 4.15: 316L stainless steel pivot hubs made by MIM for a high density disc drive application[4]. Courtesy of Parmatech Corporation, Petaluma, California, U.S.A.

made from an Fe-2% Ni sintered to a density of at least 7600 kg/m^3. This material exhibits reasonably good soft magnetic properties (B_m about 1.5 tesla and H_c about 80 A/m). The 9-pin dot matrix head is similar to that shown in Figure 4.2, with several small holes in the back plate. The 24-pin printer head has many more holes in the back plate. Specifically, there are slots between each of the 24 pins. Furthermore, several holes are located at the periphery of the part. There are both fine examples of the degree of complexity that is achievable by the metal injection molding process.

The disc drive pivot hubs shown in Figure 4.15 are made from 316L stainless steel and must be totally non-magnetic[4]. To ensure that the austenitic stainless steel stays non-magnetic, the carbon content must be maintained below 0.001%. These are shown to illustrate the degree of complexity achievable in a component fabricated by MIM. Further demands on the MIM process were that the part had to hold tight dimensional and weight tolerances. After sintering to a density of 7750 kg/m^3, the parts have a tensile strength of about 500 MN/m^2, a yield strength of 290 MN/m^2 and an elongation of 54%. The parts are used with other soft magnetic components in a positioning mechanism in the disc drive.

4.7 Summary

Pure Fe or Fe-0.45% P fabricated by powder metallurgy or metal injection molding will work reasonably well in many soft magnetic applications. These can be substituted for commercial-grade wrought steels such as AISI grades 1008, 1010, and 1020. The inherent porosity in the former materials has essentially the same degrading role as the impurities in the wrought products.

The major advantage of processing by P/M and MIM as against wrought fabrication techniques is the opportunity to obtain near-net shaped components in large quantities and precise dimensions. Sometimes, however, even these sintered products must be machined. This may include drilling, turning, milling, and tapping operations. The machinability of a P/M part can be enhanced significantly by the addition of 0.5% MnS. While aiding machinability, this additive will not degrade magnetic properties, unless the sintering temperature is raised above 1204°C (2200°F). Sintering above this temperature can promote the dissociation of MnS, leading to free S which will cause the same problems as the interstitials C, O_2, and N_2.

A second pressing or sizing operation is sometimes needed to enhance control of dimensional tolerances. This will depend on the tightness of the tolerances required, as well as the degree of shrinkage/warpage exhibited by the fabrication process. Furthermore, the supplementary pressing step may be included to cause additional densification of the part. In either case, the product should be subsequently annealed to provide the optimum soft magnetic performance.

The cleanliness of the raw ingredients, the sintering atmospheres, and the manufacturing environment play a highly pivotal role in dictating the performance level of soft magnetic components. Soft magnetic properties are far more sensitive to impurities than are mechanical properties. The goal of the parts manufacturer is to minimize impurity levels and any other defects that may have an impact on magnetic performance.

The mechanical properties of the various materials detailed in this monograph are discussed in the listed references. For a given material, a higher ductility and a lower yield strength will generally indicate improved soft magnetic properties.

The list of components and applications discussed in this monograph is by no means comprehensive. Many more cases exist that could not be covered here. Nonetheless, the examples selected were chosen because of their strong reliance on soft magnetic properties. Without the desired level of magnetic performance from these components, the assembly simply would not function. Literally, millions of soft magnetic components manufactured by powder metallurgy and metal injection molding are meeting the challenge of these demanding applications.

Further work is needed to correlate the fundamental soft magnetic properties of materials with specific applications in the field. The effect of materials and processing variables on the functional requirements of the particular device needs to be documented and communicated. Too frequently, the project engineer or magnetician resorts to prior practice in resolving the issue of which materials and fabrication techniques to utilize in a new component or assembly design. Previous designs may have worked reasonably well so that it is safe to continue with the same approach. But this practice provides no guidance on the direction as to which aspects of the material properties need to be improved for the future. An understanding of the design limits regarding material properties and component configurations would be highly desirable for the advancement of this field. The judicious implementation of Finite Element Analysis (FEA) may be very fruitful in such a program. Software predictions made by FEA need to be correlated with carefully designed experimental work.

REFERENCES

1. C. Lall and L. W. Baum, "High Performance Soft Magnetic Components by P/M and MIM", *Modern Developments in Powder Metallurgy*, Vol. 18, compiled by P.U. Gummeson and D.A. Gustafson, Metal Powder Industries Federation, Princeton, NJ, 1988, p.363-389.
2. B. T. Dodge, "New Product Applications in 400 Series Stainless Steels", APMI/MPIF Powder Metal Stainless Steel Seminar, Pittsburgh, PA, (February, 11-12, 1992).
3. C. Lall, "The Effect of Sintering Temperature and Atmosphere on the Soft Magnetic Properties of P/M Materials", *Advances in Powder Metallurgy - 1992*, Vol.3, Metal Powder Industries Federation, Princeton, NJ, 1992.
4. P. K. Johnson,"P/M Applications Diversifying Into New Markets", *Int. J. Powder Metall.*, 1992, Vol. 28, No. 3, pp. 233-241.
5. L. F. Pease, III, "Present Status of P/M Injection Molding, (MIM), - An Overview", *Progress in Powder Metall.*, 1987, Vol. 43, pp. 789-828.
6. L. F. Pease, III, "Metal Injection Molding - Overview, Process/ Industry", MIMA Symposium, (Indianapolis, IN), organized by MPIF, Princeton, NJ, (November 16, 1989).
7. P. K. Johnson,"New P/M Applications", *Int. J. Powder Metall.*, 1990, Vol. 26, No. 3, pp. 271-276.

APPENDIX A: Selected Conversion Factors used in Magnetism

The International System of Units (SI) has the following base-units:

Unit	Name	Symbol
Electric current	ampere	A
Length	meter	m
Luminous intensity	candela	cd
Mass	kilogram	kg
Thermodynamic Temperature	kelvin	K
Time	second	s

The following units relevant to magnetism are derived from these base-units as follows:

Electric charge	coulomb	$C = A \cdot s$
Electric potential	volt	$V = W \cdot A^{-1}$
Electric capacitance	farad	$F = A \cdot s \cdot V^{-1}$
Electric resistance	ohm	$\Omega = V \cdot A^{-1}$
Force	newton	$N = kg \cdot m \cdot s^{-2}$
Frequency	hertz	$Hz = s^{-1}$
Inductance	henry	$H = V \cdot s \cdot A^{-1}$
Magnetic flux	weber	$Wb = V \cdot s$
Magnetic flux density	tesla	$T = Wb \cdot m^{-2}$
Power	watt	$W = J \cdot s^{-1}$
Work, energy	joule	$J = N \cdot m$

APPENDIX A (Cont.): Selected Conversion Factors used in Magnetism

The following is a list of conversion factors that may be useful in magnetic measurements and testing.

Source: ASTM Standards Committee A-06:
ASTM Headquarters,
1916 Race Street, Philadelphia, PA 19103, USA.

Sinusoidal Waveform

Multiply	By	To Obtain
Peak current or voltage	0.70711	rms current or voltage
Peak current or voltage	0.63662	average current or voltage
Rms current or voltage	0.90032	average current or voltage

Magnetic Induction, B

Multiply	By	To Obtain
Gauss	6.4516	lines/square inch
Gauss	6.4516×10^{-8}	webers/square inch
Gauss	10^{-4}	tesla (webers/square meter)
Lines/square inch	1.5500×10^{-5}	tesla (webers/square inch)
Lines/square inch	10^{-8}	webers/square inch
Webers/square inch	1550	tesla (webers/square inch)

Magnetizing Force or Field Strength, H

Multiply	By	To Obtain
Oersted	2.0213	ampere-turns/inch
Oersted	0.79577	ampere-turns/centimeter
Oersted	79.577	ampere-turns/meter
Ampere-turns/centimeter	2.5400	ampere-turns/inch
Ampere-turns/inch	0.39370	ampere-turns/centimeter
Ampere-turns/meter	10^{-2}	ampere-turns/centimeter
Ampere-turns/meter	0.025400	amper-turns/inch

APPENDIX A (Cont.): Selected Conversion Factors used in Magnetism

Permeability, μ

Multiply	By	To Obtain
Gauss per oersted	3.1918	lines/ampere-turn inch
Gauss per oersted	3.1918×10^{-8}	webers/ampere-turn inch
Gauss per oersted	1.2566×10^{-6}	webers/ampere-turn meter
Gauss per oersted	1.2566×10^{-6}	henry/meter
Gauss per oersted	1.2566×10^{-6}	tesla meter/ampere
Webers/ampere-turn meter	2.5400×10^4	lines/ampere-turn inch
Webers/ampere-turn meter	0.025400	webers/ampere-turn inch
Webers/ampere-turn inch	10^6	lines/ampere-turn inch
Webers/ampere-turn inch	39.370	webers/ampere-turn meter
Lines/ampere-turn inch	39.370×10^{-8}	webers/ampere-turn meter
Lines/amper-turn inch	10^{-8}	webers/ampere-turn inch

Miscellaneous Conversions

Multiply	By	To Obtain
Magnetic flux (lines)	10^{-8}	webers
Henry	1.0	webers/ampere
Watts/pound	2.205	watts/kilogram
Volt amperes/pound	2.205	volt amperes per kilogram
Volume resistivity	10^{-2}	ohm meter
Energy product	7.958×10^{-3}	joules/cubic meter

APPENDIX B: Selected Magnetic Standards

Source: ASTM Standards Committee A-06 on Magnetic Testing: ASTM Headquarters, 1916 Race Street, Philadelphia, PA 19103, USA.

A 34-83   Practice for Procurement Testing and Sampling of Magnetic Materials.

A 340-87   Terminology of Symbols and Definitions Relating to Magnetic Testing.

A 341   Test Method for Direct-Current Magnetic Properties of Materials Using D-C Permeameters and the Ballistic Test Methods.

A 343-85   Test Method for Alternating-Current Magnetic Properties of Materials at Power Frequencies Using Wattmeter-Ammeter-Voltmeter Method and 25-cm Epstein Test Frame.

A 347-85   Test Method for Alternating-Current Magnetic Properties of Materials Using the Dieterly Bridge Method with 25-cm Epstein Frame.

A 596-89   Test Method for D.C. Magnetic Properties of Materials using Ring Test Procedures and the Ballistic Method.

A 677   Specification for Flat-Rolled, Nonoriented Electrical Steel, Fully Processed Type.

A 683-84   Specification for Nonoriented Electrical Steel, Semiprocessed Grades.

A 712-85   Test Method for Electrical Resistivity of Soft Magnetic Alloys.

A 753-85   Specification for Nickel-Iron Soft Magnetic Alloys.

A 773-91   Test Method for D-C Magnetic Properties of Materials Using Ring and Permeameter Procedures with D-C Hysteresigraphs.

A 801-87   Specification for Iron-Cobalt High Magnetic Saturation Alloys.

A 811-83   Specification for Soft Magnetic Iron Fabricated by Powder Metallurgy Techniques.

APPENDIX B (Cont.): Selected Magnetic Standards

A 838-85   Specification for Free-Machining Ferritic Stainless Soft Magnetic Alloys for Relay Applications.

A 839-87   Specification for Phosphorous Iron Fabricated by Powder Metallurgy Techniques.

A 848-87   Specification for Low-Carbon Magnetic Iron.

A 867-86   Specification for Iron-Silicon Relay Steels.

A 904-90   Specification for 50 Nickel - 50 Iron P/M Soft Magnetic Alloys.

APPENDIX C: Glossary of Selected Terms used in Magnetism

Source: Standard A340, ASTM Committee A-6 on Magnetic Testing: ASTM Headquarters, 1916 Race Street, Philadelphia, PA 19103, USA.

Ampere (turn) - the unit of magnetomotive force in the SI system of units. The symbol A represents the unit of electric current, ampere, in the SI system of units.

Ampere per meter, A/m - the unit of magnetizing force (field) in the SI system of units.

Anisotropic material - a material in which the magnetic properties differ in various directions.

Antiferromagnetic material - a feebly magnetic material in which almost equal magnetic moments are lined up antiparallel to each other. Its susceptibility increases as the temperature is raised until a critical (Neel) temperature is reached; above this temperature the material becomes paramagnetic.

Bloch wall - a domain wall in which the magnetic moment at any point is substantially parallel to the wall surface. Also see **domain wall.**

Coercive field or force - the d.c. magnetizing force required to restore the magnetic induction to zero after the material has been symmetrically cyclically magnetized.

Coercive field, intrinsic - the d.c. magnetizing force required to restore the intrinsic magnetic induction to zero after the material has been symmetrically cyclically magnetized.

Coercivity - the maximum value of coercive force that can be attained when the magnetic material is symmetrically cyclically magnetized to saturation induction, B_s.

Core, powder (dust) - a magnetic core comprised of small particles of electrically insulated metallic ferromagnetic material. These cores are characterized by low hysteresis and eddy-current losses.

113

APPENDIX C (Cont.): Glossary of Selected Terms used in Magnetism

Core loss, a.c., eddy current, normal - the power losses, due to eddy currents in a magnetic material that is symetrically cyclically magnetized.

Core loss, residual - the portion of the core loss power, which is not attributed to hysteresis or eddy-current losses from classical assumptions.

Core loss, a.c., specific - the active power (watts) expended per unit mass of magnetic material in which there is a cyclically varying induction of a specified maximum value, B, at a specified frequency, f.

Core plate - a generic term for any insulating material, formed metallurgically or applied externally as a thin surface coating, on sheet or strip stock used in the construction of laminated and taped wound cores.

Curie temperature - the temperature above which a ferromagnetic material becomes paramagnetic.

Current, a.c., core loss - the r.m.s. value of the in-phase component (with respect to the induced voltage) of the exciting current supplied to a coil which is linked with a ferromagnetic core.

Current, d.c. - a steady state d.c. current. A d.c. current flowing in an inductor winding will produce a uni-directional magnetomotive force in the magnetic material.

Demagnetization curve - the portion of a flux versus d.c. current plot (d.c. hysteresis loop) which lies in the second or fourth quadrant, that is, between the residual induction point, B_r, and the coercive force point, H_c. Points on this curve are designated by the coordinates, B_d and H_d.

Demagnetizing force - a magnetizing force applied in such a direction as to reduce the induction in a magnetized body.

Density - the ratio of mass to volume of a material. In c.g.s. units, g/cm^3. In SI units, kg/m^3.

Diamagnetic material - a material whose relative permeability is less than unity.

APPENDIX C (Cont.): Glossary of Selected Terms used in Magnetism

Domains, Ferromagnetic - magnetized regions, either macroscopic or microscopic in size, within ferromagnetic materials. Each domain, per se, is magnetized to intrinsic saturation at all times, and this saturation induction is unidirectional within the domain.

Domain, wall - a boundary region between two adjacent domains within which the orientation of the magnetic moment of one domain changes into a different orientation of the magnetic moment in the other domain.

Eddy current - an electric current developed in a material due to induced voltages developed in the material.

Electrical steel - a term used commercially to designate a flat-rolled iron-silicon alloy used for its magnetic properties.

Energy product - the product of the coordinate values of any point on a demagnetization curve.

Energy product, maximum - for a given demagnetization curve, the maximum value of the energy product, $(B \times M)_{max}$.

Exciting power, rms - the product of the a.c. rms exciting current and the rms voltage induced in the exciting (primary) winding on a magnetic core.

Exciting, voltage - the a.c. rms voltage across a winding linking the flux of a magnetic core. The voltage across the winding equals that across the assumed parallel combination of core inductance L_1 and core resistance R_1.

Feebly magnetic material - a material generally classified as "nonmagnetic," whose maximum normal permeability is less than 4.

Ferrimagnetic material - a material in which unequal magnetic moments are lined up antiparallel to each other. Permeabilities are of the same order of magnitude as those of ferromagnetic materials, but are lower than they would be if all atomic moments were parallel and in the same direction. Under ordinary conditions, the magnetic characteristics of ferrimagnetic materials are quite similar to those of ferromagnetic materials.

APPENDIX C (Cont.): Glossary of Selected Terms used in Magnetism

Ferrite - a term referring to magnetic oxides in general, and especially to material having the formula $M O Fe_2 O_3$, where M is a divalent metal ion or a combination of such ions. Certain ferrites, magnetically "soft" in character, are useful for core applications at radio and higher frequencies because of their advantageous magnetic properties and high volume resistivity. Other ferrites, magnetically "hard" in character, have desirable permanent magnet properties.

Ferromagnetic material - a material that, in general, exhibits that phenomena of hysteresis and saturation, and whose permeability is dependent on the magnetizing force.

Figure of merit, magnetic - the ratio of the real part of the complex relative permeability to the dissipation factor of a ferromagnetic material.

Flux linkage - the sum of all flux lines in a coil.

Flux path length - the distance along a flux loop.

Frequency, cyclic - the number of hertz (cycles/second) of a periodic quantity.

Gap length - the distance which the flux transverses in the central region of a gap in a core having an "air" (nonmagnetic) gap, where the relative permeability may be considered to be unity.

Gauss (plural gausses), G - the unit of magnetic induction in the c.g.s electromagnetic system. The gauss is equal to 1 maxwell per square centimeter or 10^{-4} tesla.

Gilbert, G_b - the unit of magnetomotive force in the c.g.s. electromagnetic system. The gilbert is a magnetomotive force of $10/4\pi$ ampere-turns.

Henry (plural henries) - the unit of self or mutual inductance. The henry is the inductance of a circuit in which a voltage of one volt (V) is induced by a uniform rate of change one ampere per second (A/s) in the circuit. Alternatively, it is the inductance of a circuit in which an electric current of one ampere per second produces a flux linkage of one weber turn (Wb turn) or 10^8 maxwell-turns.

APPENDIX C (Cont.): Glossary of Selected Terms used in Magnetism

Hysteresis loop, normal - a closed curve obtained with a ferromagnetic material by plotting (usually to rectangular coordinates) corresponding d.c. values of magnetic induction (B) for ordinates and magnetizing force (H) for abscissa when the material is passing through a complete cycle between equal definite limits of either magnetizing force +/- H_m or magnetic induction, +/-B_m. In general the normal hysteresis loop has mirror symmetry with respect to the origin of the B and H axes but this may not be true for special materials.

Hysteresis loop, loss - the power expended in a single slow excursion around a normal hysteresis loop. The energy is the integrated area enclosed by the loop measured in gauss-oersteds.

Inductance, core - the effective parallel circuit inductance of a ferromagnetic core based upon a hypothetical nonresistive path that is exclusively considered to carry the magnetizing current.

Inductance, intrinsic (ferric) - that portion of the self-inductance which is due to the intrinsic induction in a ferromagnetic core.

Inductance, mutual - the common property of two electrical circuits that determines the flux linkage in one circuit (the secondary) produced by a given current in the other circuit (primary).

Inductance, self - that property of an electric circuit which determines the flux linkage produced by given current in the circuit.

Induction, intrinsic - the vector difference between the d.c. magnetic induction in a magnetic material and the magnetic induction that would exist in a vacuum under the influence of the same magnetizing force.

Induction, maximum:
(1) B_m - the maximum value of induction, B, in a d.c. hysteresis loop. The tip of this loop has the magnetostatic coordinates H_m, B_m, which exist simultaneously.
(2) B_{max} - the maximum value of induction, B, in an a.c. flux-current loop.

Induction, remanent - the magnetic induction that remains in a magnetic circuit after the removal of an applied magnetomotive force.

APPENDIX C (Cont.): Glossary of Selected Terms used in Magnetism

Induction, residual - the value of magnetic induction corresponding to zero magnetizing force when the magetic material is subjected to symmetrically cyclically magnetized conditions.

Induction, saturation - the maximum intrinsic induction possible in a material.

Induction curve, intrinsic (ferric) - a curve of a previously demagnetized specimen depicting the relation between intrinsic induction and corresponding ascending values of magnetizing force.

Induction curve, normal - a curve of a previously demagnetized specimen depicting the relation between normal induction and corresponding ascending values of magnetizing force. This curve starts at the origin of B and H.

Insulation resistance - the apparent resistance between adjacent contacting laminations, calculated as a ratio of the applied voltage to conduction current. This parameter is normally a function of the applied force and voltage.

Isotropic material - material in which the magnetic properties are the same for all directions.

Leakage flux - the flux outside the boundary of the practical magnetic circuit.

Magnet - a body that produces a magnetic field external to itself.

Magnetic aging - the change in the magnetic properties of a material resulting from a metallurgical change due to a normal or specified aging condition.

Magnetic circuit - a region at whose surface the magnetic induction is tangential.

APPENDIX C (Cont.): Glossary of Selected Terms used in Magnetism

Magnetic field of induction - the magnetic flux field induced in a region such that a conductor carrying a current in the region would be subjected to a mechanical force, and an electromotive force would be induced in an elementary loop rotated with respect to the field in such a manner as to change the flux linkage.

Magnetic field strength - an alternate term for magnetizing force.

Magnetic flux - the product of the magnetic induction, B, and the area of a surface (or cross-section), A, when the magnetic induction B is uniformly distributed and normal to the plane of the surface.

Magnetic flux density - an alternate term for magnetic induction.

Magnetic induction (flux density) - that magnetic vector quantity which at any point in a magnetic field is measured either by the mechanical force experienced by an element of electric current at the point, or by the electromotive force induced in an elementary loop during any change in flux linkages with the loop at the point.

Magnetic line of force - an imaginary line in a magnetic field which at every point has the direction of magnetic induction at that point.

Magnetic moment - a measure of the magnetizing force, H, produced at points in space by a plane current loop or a magnetized body.

Magnetic particle inspection method - a method for detecting magnetic discontinuities or inhomogeneities on or near the surface in suitably magnetized materials, that employs finely divided magnetic particles that tend to congregate in regions of magnetic nonuniformity associated with the magnetic discontinuities or inhomogeneities.

Magnetic polarization, J - in the c.g.s.-e.m.u. system, the intrinsic induction divided by 4π is sometimes called magnetic polarization or magnetic dipole moment per unit volume.

Magnetic pole - the magnetic poles of a magnet are those portions of the magnet toward which or from which the external magnetic induction appears to converge or diverge respectively.

APPENDIX C (Cont.): Glossary of Selected Terms used in Magnetism

Magnetic pole strength - the magnetic moment divided by the distance between the poles.

Magnetizing force, maximum:
   (a) H_m - the maximum value of H in a d.c. hysteresis loop.
   (b) H_{max} - the maximum value of H in an a.c. flux-current loop.

Magnetomotive force - the line integral of the magnetizing force around any flux loop in space.

Magnetostatic - the magnetic condition when the values of magnetizing force and induction are considered to remain invariant with time during the period of measurement. This is often referred to as a d.c. condition.

Magnetostriction - the change in dimensions of a body resulting from magnetization.

Mass, total - the actual mass of a magnetic core.

Maxwell - the unit of magnetic flux in the c.g.s-e.m.u electromagnetic system. One maxwell equals 10^{-8} weber.

Nonmagnetic - a relative term describing a material which, for practical purposes, may be considered to have a relative permeability close to unity.

Nonoriented electrical steel - a flat-rolled electrical steel which has approximately the same magnetic properties in all directions.

Oersted - the unit of magnetizing force (magnetic field strength) in the c.g.s-e.m.u electromagnetic system. One oersted equals a magnetomotive force of 1 gilbert/cm of flux path. One oersted equals $1000/4\pi$ or 79.58 ampere-turns per meter.

Paramagnetic material - a material having a relative permeability which is slightly greater than unity, and which is practically independent of the magnetizing force.

APPENDIX C (Cont.): Glossary of Selected Terms used in Magnetism

Permeability, maximum - the highest value of permeability achieved when the magnetic material is subject to a symmetrically cyclically magnetized condition.

Permeability, relative - the ratio of the absolute permeability of a material of the magnetic constant, giving a pure numeric parameter.

Permeability, unoccupied space - the permeability of free space (vacuum), identical with the magnetic constant.

Permeance - the reciprocal of the reluctance of a magnetic circuit.

Power factor, magnetic - (a) the cosine of the angle between vectors representing the rms values of the applied voltage of a circuit and the current in circuit. (b) the ratio of the active (real) power to the apparent power in an a.c. circuit.

Power active (real) - the product of the rms current in a circuit, the rms voltage across the circuit and the cosine of the angular phase difference between the current and voltage.

Reluctance - that quantity which determines the magnetic flux, resulting from a given magnetomotive force, around a magnetic circuit.

Reluctivity - the reciprocal of the permeability of a medium.

Remanence - the maximum value of the remanent induction for a given geometry of the magnetic circuit.

Resistance, core - the effective a.c. resistance of a hypothetical parallel resister that is considered to carry exclusively the core loss current, I_c, when a voltage is applied to the terminals of a coil encircling a magnetic core.

Resistance, winding - the effective a.c. series resistance of an inductor when no ferromagnetic materials are present.

Retentivity - the property of a magnetic material which is measured by its maximum value of the residual induction.

APPENDIX C (Cont.): Glossary of Selected Terms used in Magnetism

SI - an abbreviation for the International System of Units.

Skin effect, magnetic - the nonuniform magnetodynamic term applies to the nonuniform distribution of induction existing at various points in the cross section of a magnetic core. Skin effect is produced, primarily by eddy-current phenomena and it increases with the frequency of a.c. excitation. It can ordinarily be neglected in testing at commercial power frequencies.

Susceptibility - a ratio of the intrinsic induction, B_i, due to the magnetization of a material to the induction in space due to the influence of the corresponding magnetizing force, H.

Susceptibility, initial - the limiting value of susceptibility when the intrinsic induction approaches zero.

Susceptibility, mass - the susceptibility divided by the density of a body is called the susceptibility per unit mass, x, or simply the mass susceptibility.

Tesla, T - the SI unit of magnetic induction. One tesla is equal to 1.0 Wb/m^2 or 10^4 gausses.

Volt-ampere - the unit of apparent power.

Watt - the unit of active power. One watt is energy, work, or quantity of heat expended at a rate of one joule per second.

Weber - the unit of magnetic flux. The weber is the magnetic flux whose decrease to zero when linked with a single turn, induces in the turn a voltage whose time integral is one volt-second. One weber equals 10^8 maxwells.

Winding loss, (copper loss) - the power expended, as heat, in the conductors of an inductor or resistor, or both, due to the electric current in them.

APPENDIX D: Magnetic Properties of Selected Materials (expressed in c.g.s units)

Table 3.1C Data Supplied with ASTM Standard Toroids (1979)

	TOROID NUMBER		
	67004	67007	67010
MATERIAL:	GRAIN OR. M-5 Fe-3% Si	NON-OR. M-36 Fe-1.4% Si	NON-OR. M-15 Fe-4% Si
DENSITY (g/cm^3)	7.65	7.75	7.65
MAG. PATH L (cm)	35.91	35.91	35.91
ACTIVE WT (g:lbs)	414:0.9127	428:0.9436	414:0.9127
SAMPLE AREA (cm^2)	1.507	1.538	1.507
THICKNESS (in)	0.012	0.0185	0.014
PRIMARY TURNS	265	265	265
SECONDARY TURNS	265	265	265
PRIMARY WINDING RESISTANCE (ohm)	0.371	0.371	0.370
FLUX VOLTS AT 10 GAUSS, 60 Hz	10.64	10.86	10.64

APPENDIX D (Cont.): Magnetic Properties of Selected Materials (expressed in c.g.s units)

Table 3.2C Measured Data For ASTM Toroids
              (after Lall and Baum[1])

	TOROID NUMBER		
	67004	67007	67010
MATERIAL:	GRAIN OR. M-5 Fe-3% Si	NON-OR. M-36 Fe-1.4% Si	NON-OR. M-15 Fe-4% Si

ASTM DATA AT H = 10 Oe (1.078 Amps)

Average of all B_m values, kgauss	18.423	15.773	14.182

AT CALCULATED H = 9.27 Oe (1.0 Amp)

B_m (kG)	18.13	15.50	14.25
B_r (kG)	15.88	14.00	10.75
H_c (Oe)	0.14	0.55	0.32
μ_m	48,000	12,000	13,000

AT CALCULATED H = 15 Oe (1.6 Amps)

B_m (kG)	18.50	15.88	14.63
B_r (kG)	15.88	14.06	10.75
H_c (Oe)	0.15	0.55	0.32
μ_m	45,000	11,000	13,000

AT CALCULATED H = 19 Oe (2 Amps)

B_m (kG)	18.86	16.00	14.75
B_r (kG)	15.88	14.06	10.75
H_c (Oe)	0.15	0.56	0.32
μ_m	40,000	11,000	13,000

APPENDIX D (Cont.): Magnetic Properties of Selected Materials (expressed in c.g.s units)

Table 3.3C Showing Effect of Different Applied Fields (after Lall[20]).

SOURCE	PROPERTY				
	DENSITY	B_m	B_r	μ_m	H_c
	(g/cm^3)	(kG)	(kG)		(Oe)
Pure iron:					
At H = 15 Oe	7.0	10.9	9.7	2400	1.70
At H = 25 Oe	7.0	11.5	10.0		1.70
At H = 100 Oe	7.0	13.5	11.0		1.70
Fe-0.45% P:					
At H = 15 Oe	7.0	11.9	11.2	3300	1.60
At H = 25 Oe	7.0	12.4	11.4		1.60
At H = 100 Oe	7.0	14.3	12.4		1.60
Fe-50% Ni:					
At H = 15 Oe	7.2	9.9	8.1	12,300	0.28
At H = 25 Oe	7.2	10.1	8.4		0.28
At H = 100 Oe	7.2	11.4	8.8		0.28

NOTE: All samples sintered at 2280°F (1249°C), 45 min., in DA.

APPENDIX D (Cont.): Magnetic Properties of Selected Materials (expressed in c.g.s units)

Table 3.4C Magnetic Data on Pure Iron

SOURCE	PROPERTY				
	DENSITY	B_m	B_r	μ_m	H_c
	(g/cm^3)	(kG)	(kG)		(Oe)
WROUGHT					
ASTM A 848	7.86	17.0	12.8	5,000	1.0

(Electrical resistivity 13 micro-ohm·cm; Core loss 2700 erg/cm^3).

POWDER METAL					
1. ASTM A811-87	6.6	9.0	7.6	1800	2.2
H = 15 Oe	6.9	10.6	9.1	2100	2.0
	7.2	12.3	10.7	2500	2.0
2. Lall & Baum[1]: H = 25 Oe					
2300°F, DA	6.8	11.4	9.6	2900	1.65
H=25 Oe, 45 min.	7.2	13.6	11.0	3700	1.60
	7.4	14.7	12.9	4700	1.50

(Resistivity 14, 12, 11 micro-ohm·cm at 6.8, 7.2, and 7.4 g/cm^3, respectively).

3. McDermott[21]: H = 15 Oe					
2050°F, DA	6.7	9.8	8.3	1990	2.02
30 min.	7.0	11.4	9.7	2320	2.07
	7.2	12.4	10.6	2650	2.08
4. McDermott[21]: H = 15 Oe					
2050°F, H2	6.8	10.6	9.2	2530	1.84
30 min.	7.1	12.1	10.7	3070	1.84
	7.3	12.9	11.6	3340	1.84
5. McDermott[21]: H = 15 Oe					
2300°F, DA	6.7	10.0	7.7	1820	2.05
30 min.	7.0	11.7	9.1	2180	2.05
	7.3	12.7	10.5	2790	1.95

APPENDIX D (Cont.): Magnetic Properties of Selected Materials (expressed in c.g.s units)

Table 3.4C (Cont.) Magnetic Data on Pure Iron

SOURCE	PROPERTY				
	DENSITY	B_m	B_r	μ_m	H_c
	(g/cm^3)	(kG)	(kG)		(Oe)
6. McDermott[21]: H = 15 Oe					
2300°F, H$_2$	6.9	11.0	9.5	2730	1.76
30 min.	7.1	12.5	10.9	3190	1.72
	7.3	13.3	11.7	3570	1.68
7. Gagne et al.[22]: H = 15 Oe					
1120°C, H$_2$	7.4	12.4	11.6	2900	1.7
1260°C, H$_2$	7.4	12.5	12.0	3600	1.5
8. Lall[20]: H = 25 Oe					
2300°F, vac.	7.0	11.7	10.9	3500	1.4
9. Moyer[23]: H = 15 Oe					
2050°F, H$_2$ + N$_2$	7.0	11.4	11.0	2800	1.8
Shot-peened	7.0	9.7	8.7	1700	2.0
10. Mossner[25]: H = 15 Oe					
2050°F,	7.0	10.0	8.9	2000	2.0
2350°F	7.0	10.7	9.4	2400	1.6

APPENDIX D (Cont.): Magnetic Properties of Selected Materials (expressed in c.g.s units)

Table 3.5C Magnetic Data On Fe-0.45% P

SOURCE	PROPERTY				
	DENSITY	B_m	B_r	μ_m	H_c
	(g/cm^3)	(kG)	(kG)		(Oe)

WROUGHT
NOT AVAILABLE IN WROUGHT FORM.

POWDER METAL

1. ASTM A-839	6.8	10.7	8.7	2400	1.7
	7.1	11.9	9.9	2800	1.7
	7.2	12.7	10.8	3100	1.6

2. Lall and Baum[1]: H = 25 Oe

2300°F, DA	7.0	12.3	9.9		1.2
45 min.	7.2	13.4	11.2	4800	1.00
	7.4	14.6	12.6		0.75

(Resistivity 23, 21, 20 micro-ohm·cm at 7.0, 7.2, and 7.4 g/cm^3, respectively).

3. McDermott[21]: H = 15 Oe

2050°F, DA	6.84	11.1	9.7	2830	1.75
30 min.	7.12	12.4	11.0	3290	1.75
	7.27	13.2	11.8	3700	1.71

4. McDermott[21]: H = 15 Oe

2300°F, DA	7.15	12.7	11.0	4220	1.38
30 min.	7.35	13.9	13.5	4070	1.33
	7.45	13.9	11.8	4450	1.29

5. Gagne et al.[22]: H = 15 Oe

1120°C, H_2 30 min.	7.4	12.6	12.3	4300	1.4
1260°C, H_2 30 min.	7.4	13.2	12.8	5000	1.2

APPENDIX D (Cont.): Magnetic Properties of Selected Materials (expressed in c.g.s units)

Table 3.5C (Cont.) Magnetic Data On Fe-0.45% P

SOURCE	PROPERTY:				
	DENSITY	B_m	B_r	μ_m	H_c
	(g/cm^3)	(kG)	(kG)		(Oe)
6. Moyer[23]: H = 15 Oe					
2050°F	7.2	12.2	11.0	3200	1.5
2350°F	7.2	13.2	11.8	4800	1.3
7. Lall[20]: H = 25 Oe					
2300°F, vac.	7.2	13.5	12.5	6000	1.0
8. Moyer[23]: H = 15 Oe					
2050°F, $H_2 + N_2$	7.1	12.2	11.9	3600	1.5
Shot-peened	7.1	9.9	8.7	2000	1.9
9. Mossner[25]: H = 15 Oe					
2050°F,	7.2	10.2	8.7	2250	1.8
2350°F	7.4	13.1	11.5	5550	0.9

APPENDIX D (Cont.): Magnetic Properties of Selected Materials (expressed in c.g.s units)

Table 3.6C Magnetic Data On Fe-3% Si

SOURCE	PROPERTY				
	DENSITY	B_m	B_r	μ_m	H_c
	(g/cm^3)	(kG)	(kG)		(Oe)

WROUGHT

ASM Handbook[2]   7.65   20.1 (B_s)   -   8,000   0.7
(3% Si steel has electrical resistivity of 47 mico-ohm·cm).

Carpenter Technology[27]   7.65   16   6   5,000   0.7
(2.5% Si steel, "B-FM" grade, B_m at H = 10 Oe, B_s = 20.6 kG, resistivity 40 micro-ohm·cm)

Carpenter Technology[27]   7.60   15   4   4,000   0.6
(4% Si steel, "C" grade, B_m at H = 10 Oe, B_s = 20 kG, resistivity 58 micro-ohm·cm)

POWDER METAL

1. Lall and Baum[1]: H = 25 Oe

2300°F	6.8	11.7	9.4	2900	1.30
45 min.	7.0	13.1	10.9	3700	1.15
	7.2	13.9	11.8	4900	1.00

(Resistivity 59, 55, 52 micro-ohm·cm at 6.8, 7.0, and 7.2 g/cm^3, respectively).

2. Lall[20]: H = 25 Oe

2300°F, vac.	7.2	12.8	11.7	7000	0.8

3. Mossner[25]: H = 15 Oe

2050°F,	6.7	9.9	6.8	1750	1.8
2350°F	7.4	11.9	9.1	4050	0.9

4. Moyer and Ryan[26]: H = 15 Oe

2250°F, vac. 1 hr.	7.2	13.5	12.1	7500	0.9
2250°F, vac. 1 hr.	7.4	12.3	7.5	7000	0.6

APPENDIX D (Cont.): Magnetic Properties of Selected Materials (expressed in c.g.s units)

Table 3.7C	Magnetic Data On Fe-50% Ni				
SOURCE			**PROPERTY:**		
	DENSITY	B_m	B_r	μ_m	H_c
	(g/cm^3)	(kG)	(kG)		(Oe)

WROUGHT

ASM Handbook[2]   8.2   16.0   8.0   70,000   0.05

("hypernik" has an electrical resistivity of 50 micro-ohm·cm).

POWDER METAL

1. Lall and Baum[1]: H = 25 Oe

2300°F, DA	6.8	9.3	7.1		0.26
45 min.	7.1	10.9	8.0		0.25
	7.5	12.7	9.4	21000	0.24

(Resistivity 78, 69, 60 micro-ohm·cm at 6.8, 7.1, and 7.5 g/cm^3, respectively).

2. Lall[20]: H = 25 Oe

2300°F, vac	7.4	10.8	8.6	10600	0.3

3. Mossner[25]: H = 15 Oe

2050°F	7.1	11.5	8.5	11000	0.4
2350°F	7.3	12.9	9.0	16000	0.3

4. Moyer and Ryan[26]: H = 15 Oe

2275°F, vac. 2 hr.	7.3	11.2	7.0		0.3
2275°F, vac. 2 hr.	7.5	12.3	7.5		0.3

APPENDIX D (Cont.): Magnetic Properties of Selected Materials (expressed in c.g.s units)

Table 3.8C Magnetic Data On Ferritic Stainless Steels

SOURCE	PROPERTY:				
	DENSITY	B_m	B_r	μ_m	H_c
	(g/cm^3)	(kG)	(kG)		(Oe)

WROUGHT

Carpenter Technology[27]	7.62	12	6	2000	2.0

(430F, B_m at H = 10 Oe, B_s=14.2 kG)

POWDER METAL

410L

1. Lall[20]: H = 25 Oe

2300°F, Vac. 45 min.	7.1	10.9	9.4	2200	2.0
2350°F, H$_2$	7.1	5.6	–	320	7.4

430L

2. McDermott[21]: H = 15 Oe

2050°F, H$_2$	6.45	7.28	4.70	1000	2.29
30 min.	6.67	7.90	5.09	1043	2.32

3. McDermott[21]: H = 30 Oe

2050°F, DA	5.81	0.34	0.03	11	2.9
30 min.	6.13	0.41	0.05	13	3.7
	6.42	0.45	0.05	14	3.5

4. Moyer and Jones[13]: H = 15 Oe

2300°F, H$_2$	7.25	10.5	8.0	1900	1.8

5. Svilar and Ambs[14]: (H = 15 Oe, vac + backfill of listed gas)

2050°F, H$_2$	6.69	8.1	7.4	1200	2.6
2250°F, H$_2$	7.11	9.8	8.9	1800	2.1
2050°F, H$_2$ + N$_2$	6.49	0.16	0.09		9.3
2250°F, H$_2$ + N$_2$	6.99	0.38	0.24		8.4

APPENDIX D (Cont.): Magnetic Properties of Selected Materials (expressed in c.g.s units)

Table 3.8C (Cont.) Magnetic Data On Ferritic Stainless Steels

SOURCE	PROPERTY:				
	DENSITY	B_m	B_r	μ_m	H_c
	(g/cm^3)	(kG)	(kG)		(Oe)
434L					
6. Lall[20]: H = 25 Oe					
2300°F, Vac. 45 min.	7.0	10.1	8.4	1700	2.0
2350°F, H$_2$, 45 min.	7.1	6.5	–	450	2.8
7. McDermott[21]: H = 15 Oe					
2350°F, DA, 30 min.	6.43	7.28	4.63	1092	2.01
2350°F, H$_2$, 30 min.	6.65	7.91	4.83	1165	1.90
8. McDermott[21]: H = 15 Oe					
2050°F, DA	5.83	0.53	0.09	416.8	5.2
30 min.	6.03	0.63	0.13	719.7	6.0
	6.29	0.79	0.19	424.9	6.7
9. Mossner[25]: H = 15 Oe					
2350°F	7.1	8.9	4.4	1275	1.5
10. Moyer and Jones[13]: H = 15 Oe					
2250°F, H$_2$	7.35	9.7	7.7	1600	1.8

APPENDIX D (Cont.): Magnetic Properties of Selected Materials (expressed in c.g.s units)

Table 3.9C Magnetic Data on Special P/M Alloys

SOURCE	PROPERTY:				
	DENSITY	B_m	B_r	μ_m	H_c
	(g/cm^3)	(kG)	(kG)		(Oe)
Fe-81%Ni-2%Mo; Permaloy					
1. Lall and Baum[1]: H = 25 Oe					
2300°F, DA, 45 m.	7.8	7.2	4.8	77000	1.07
Fe-50%Co-2%V; Permendur					
2. Lall and Baum[1]: H = 25 Oe					
2300°F, DA, 45 min.	7.2	12.4	7.3		2.2
	7.4	14.7	12.9	4700	1.50
3. Takeda et al.[16]: H = 100 Oe					
1260°C, H$_2$, 3 hrs.	8.0	23.0	–	5700	1.20
Fe-0.8% P					
4. Lall and Baum[1]: H = 25 Oe					
2300°F, DA	7.0	12.7	10.8		1.48
45 min.	7.2	13.2	11.3		1.49
	7.4	14.2	11.5		0.87
(Resistivity 32, 30, 28 micro-ohm·cm at 7.0, 7.2, and 7.4 g/cm^3, respectively).					
5. Gagne et al.[23]: H = 15 Oe					
1120°C	7.4	12.9	12.5	5200	1.2
1260°C	7.5	13.5	13.2	6000	1.0
6. Moyer[24]: H = 15 Oe					
2050°F, H$_2$ + N$_2$	7.1	12.3	12.0	4100	1.45
Shot-peened	7.1	9.8	8.3	2100	1.8
7. Mossner[25]: H = 15 Oe					
2050°F	7.2	12.8	11.8	5300	1.2
2350°F	7.4	14.2	13.0	9100	0.36

APPENDIX D (Cont.): Magnetic Properties of Selected Materials (expressed in c.g.s units)

Table 3.10C Magnetic Data On Metal Injection Molded Alloys

SOURCE	PROPERTY:					
	DENSITY		B_m	B_r	μ_m	H_c
	(g/cm^3)		(kG)	(kG)		(Oe)
After Lall and Baum[1]						
Fe-2% Ni, 2400°F	7.67	97%	15.1	12.9		1.03
Fe-50% Ni, 2400°F	7.66	94%	12.7	4.2		0.20
Fe-3% Si, 2400°F	7.55	98%	15.0	12.1		0.57
Then, H$_2$ Annealed	7.55	98%	14.7	11.4		0.59
Fe-3% Si, 2250°F	7.54	98%	14.4	6.2		0.75
Fe-6% Si, 2400°F	7.54	99%	13.2	11.2		0.58
Then, H$_2$ Annealed	7.41	98%	13.7	12.2		0.58
After Baum[24]						
Fe, 2500°F, DA	7.60	97%	15.3	13.7		3.7
Fe, 2500°F, DA	7.55	96%	15.5	13.4		2.3
Fe-3% Si, 2400°F, DA	7.55	98%	14.5	10.7		0.72
Fe-3% Si, 2400°F, DA	7.55	98%	15.0	12.1		0.64
Fe-6% Si, 2400°F, DA	7.42	99%	13.7	12.1		0.50
Fe-50% Ni, 2400°F, DA	7.66	93%	12.7	4.2		0.20
430L SS, 2500°F, VAC	7.40	95%	11.5	5.4		2.54
After Mossner[25]						
Fe-3% Si			14.8	8.8		0.78
			14.6	11.0		0.7
Fe-50% Ni			12.2	5.3		0.20
			10.4	6.4		0.24
430L SS			8.5	2.2		1.5

APPENDIX D (Cont.): Magnetic Properties of Selected Materials (expressed in c.g.s units)

Table 3.11C Effect of Alloying Elements on Magnetic Properties of Iron

SOURCE	PROPERTY				
	DENSITY	B_m	B_r	μ_m	H_c
	(g/cm^3)	(kG)	(kG)		(Oe)
POWDER METAL:					
Pure Fe					
1. Lall and Baum[1]: H = 25 Oe					
2300°F, DA	6.8	11.4	9.6	2900	1.65
45 min.	7.2	13.6	11.0	3700	1.60
	7.4	14.7	12.9	4700	1.50
2. Lall[20]: H = 25 Oe					
2280°F, DA, 50 min.	7.0	11.5	10.8	2400	1.70
Fe-0.5% C					
3. Lall[20]: H = 25 Oe					
2280°F, DA, 50 min.	7.0	10.8	9.5		4.8
Fe-2%Cu-0.8%C					
4. Lall[20]: H = 25 Oe					
2280°F, DA, 50 min.	7.0	19.5	8		7.5

Table 3.12C   Comparison of Materials Produced by Wrought and Powder Metallurgy Techniques

	Fe	Fe-0.45% P	Fe-3% Si	Fe-50% Ni
MAXIMUM INDUCTION, B_m (kG)				
P/M	14	14	13	12
WROUGHT	17	–	16	15
RESIDUAL INDUCTION, B_r (kG)				
P/M	12	12	11	9
WROUGHT	13	–	7	8
COERCIVE FIELD, H_c (Oe)				
P/M	1.6	1.0	1.0	0.2
WROUGHT	1.0	–	0.8	0.1
MAXIMUM PERMEABILITY, μ_m				
P/M	4,000	5,000	5,000	20,000
WROUGHT	5,000	–	7,000	70,000
ELECTRICAL RESISTIVITY (micro-ohm·cm)				
P/M	12	20	55	70
WROUGHT	10	–	50	50
CORE LOSS (watts/kg)				
P/M	24	16	5	–
WROUGHT	12	–	4	–
DENSITY of P/M				
(g/cm^3)	7.2	7.2	7.2	7.4
% OF WROUGHT	92	92	94	90

INDEX

Hot isostatic pressing, 32
Hysteresis, 10-11, 14, 25
Hysteresis graphs, 11, 14, 19, 21
Hysteresis loss, 16

I
Initial permeability, 10
Intensity of magnetization, 9
Intrinsic coercive field, 21
Intrinsic curve, 21
Intrinsic demagnetization, 21
Iron
    effect of alloying elements on
        magnetic properties, 83, 136
    soft magnetic properties of,
        53-59, 126-127
Iron-nickel, soft magnetic
        properties of, 67-70, 131
Iron-phosphorus, soft magnetic
        properties of, 59-63, 128-129
Iron-silicon, soft magnetic
        properties of, 64-67, 130

L
Leakage factor, 22
Lenz's Law, 16
Linear voice coil motors, 100-102

M
Magnetic aging, 82
Magnetic anisotrophy, 12
Magnetic compass needle, 2
Magnetic dipole, 8
Magnetic domain rotation, 14
Magnetic domains, 12
Magnetic field, 1, 4
    external, 12
    internal-molecular, 12
Magnetic flux, 24
Magnetic force, 3
Magnetic hysteresis curve,
    10-11, 14

Magnetic induction, 8
Magnetic pole, 4
Magnetic properties in c.g.s. units,
    123-137
Magnetic standards, 111-112
Magnetic testing, 50-51
Magnetism, measurement of, 24
Magnetization, 9
Magnetometer, vibrating sample, 24
Magnetostriction, 15
Magnets
    hard, 20
    permanent, 20
    soft, 20
    temporary, 20
Material transport methods, 36-37
Maximum energy product, 21
Maximum induction, 11
Maximum permeability, 11
Measurement of magnetism, 24
Metal injection molding, 30, 40-41
    binder systems, 44-45
    soft magnetic examples formed
        by, 102-104
Metal injection molding
        components, soft magnetic
        properties of, 77-78, 135
Molecular field theory, 12
Moving charge, 3

N
Neck ratio, 36
Newton's Second Law of Motion, 2
Newton's Universal Law of
    Gravitation, 2

O
Oersted, H.C., 4
Open magnetic circuit, 17

P
Paramagnetism, 10, 26

NOTES